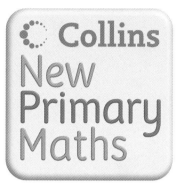

Collins New Primary Maths

Assessment Pack 6

Author: Peter Clarke

William Collins' dream of knowledge for all began with the publication of his first book in 1819. A self-educated mill worker, he not only enriched millions of lives, but also founded a flourishing publishing house. Today, staying true to this spirit, Collins books are packed with inspiration, innovation and practical expertise. They place you at the centre of a world of possibility and give you exactly what you need to explore it.

Collins. Freedom to Teach.

Published by Collins
An imprint of HarperCollinsPublishers
77 – 85 Fulham Palace Road
Hammersmith
London
W6 8JB

Browse the complete Collins catalogue at
www.collinseducation.com

ISBN-13 978 0 00 722053 3

British Library Cataloguing in Publication Data
A Catalogue record for this publication is available from the British Library

Cover design by Laing&Carroll
Cover artwork by Jonatronix Ltd
Internal design and page make-up by Neil Adams
Illustrations by Neil Adams and Bridget Dowty
Edited by Ros Davies

Printed and bound by Martins the Printers, Berwick-upon-Tweed

Contents

Appendix A: Record-keeping formats

169

Appendix B: Resource Copymasters to accompany Adult Directed Tasks

186

Introduction

What does the Primary National Strategy (PNS) *Renewed Framework for Mathematics* (2006) say about assessment?

The PNS *Renewed Framework for Mathematics* identifies two main purposes of assessment:

- Assessment *for* learning (formative on-going assessment)
- Assessment *of* learning (summative assessment)

Assessment *for* learning involves both pupils and teachers finding out about the specific strengths and weaknesses of individual children, and the class as a whole, and using this to inform future teaching and learning.

Assessment *for* learning:

- is part of the planning process
- is informed by learning objectives
- engages children in the assessment process
- recognises the achievements of all children
- takes account of how children learn
- motivates learners.

Assessment *of* learning is any assessment that summarises where individual children, and the class as a whole, are at a given point in time. It provides a snapshot of what has been learned.

The *Collins NEW Primary Maths (CNPM) Assessment Packs*

The *CNPM Assessment Packs* aim to provide guidance in both Assessment *for* learning and Assessment *of* learning.

The *CNPM Assessment Packs* consist of three key features:

- Section 1: Adult Directed Tasks
- Section 2: Pupil Self assessments
- Section 3: Tests

Section 1: Adult Directed Tasks

Purposes

- To assist in identifying particular children's strengths and weaknesses.
- To inform future planning and teaching of individual children and the class as a whole.
- To provide some guidance about what to do for those children who are achieving above or below expectations.

When to use this feature

- Anytime throughout the year when you are uncertain about a child's, or a group of children's, understanding of a particular objective.

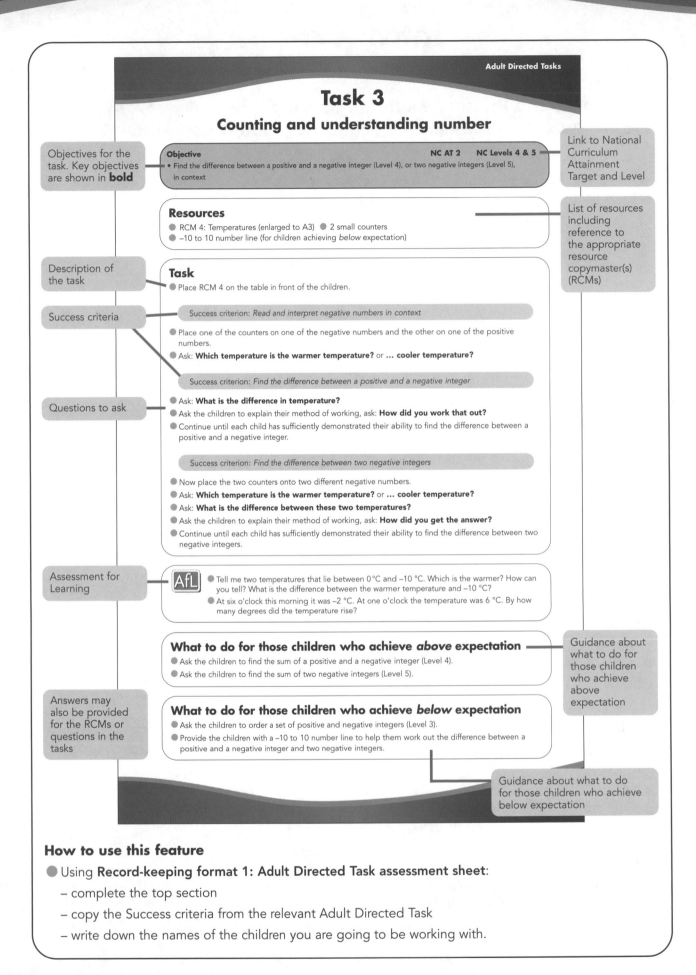

Adult Directed Tasks

Task 3
Counting and understanding number

Objectives for the task. Key objectives are shown in **bold**

Objective   NC AT 2   NC Levels 4 & 5
• Find the difference between a positive and a negative integer (Level 4), or two negative integers (Level 5), in context

Link to National Curriculum Attainment Target and Level

Resources
● RCM 4: Temperatures (enlarged to A3)   ● 2 small counters
● –10 to 10 number line (for children achieving *below* expectation)

List of resources including reference to the appropriate resource copymaster(s) (RCMs)

Description of the task

Task
● Place RCM 4 on the table in front of the children.

Success criteria

Success criterion: *Read and interpret negative numbers in context*

● Place one of the counters on one of the negative numbers and the other on one of the positive numbers.
● Ask: **Which temperature is the warmer temperature?** or **... cooler temperature?**

Success criterion: *Find the difference between a positive and a negative integer*

Questions to ask

● Ask: **What is the difference in temperature?**
● Ask the children to explain their method of working, ask: **How did you work that out?**
● Continue until each child has sufficiently demonstrated their ability to find the difference between a positive and a negative integer.

Success criterion: *Find the difference between two negative integers*

● Now place the two counters onto two different negative numbers.
● Ask: **Which temperature is the warmer temperature?** or **... cooler temperature?**
● Ask: **What is the difference between these two temperatures?**
● Ask the children to explain their method of working, ask: **How did you get the answer?**
● Continue until each child has sufficiently demonstrated their ability to find the difference between two negative integers.

Assessment for Learning

AfL ● Tell me two temperatures that lie between 0 °C and –10 °C. Which is the warmer? How can you tell? What is the difference between the warmer temperature and –10 °C?
● At six o'clock this morning it was –2 °C. At one o'clock the temperature was 6 °C. By how many degrees did the temperature rise?

What to do for those children who achieve *above* expectation
● Ask the children to find the sum of a positive and a negative integer (Level 4).
● Ask the children to find the sum of two negative integers (Level 5).

Guidance about what to do for those children who achieve above expectation

Answers may also be provided for the RCMs or questions in the tasks

What to do for those children who achieve *below* expectation
● Ask the children to order a set of positive and negative integers (Level 3).
● Provide the children with a –10 to 10 number line to help them work out the difference between a positive and a negative integer and two negative integers.

Guidance about what to do for those children who achieve below expectation

How to use this feature

● Using **Record-keeping format 1: Adult Directed Task assessment sheet:**

– complete the top section

– copy the Success criteria from the relevant Adult Directed Task

– write down the names of the children you are going to be working with.

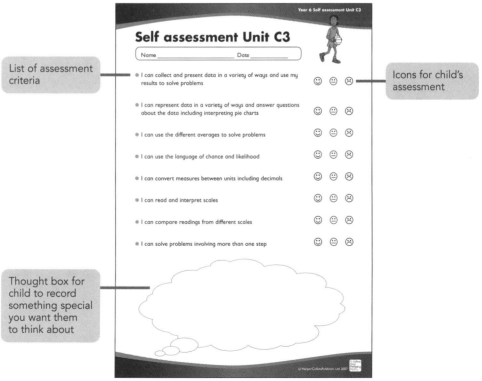

Record-keeping format 1 Adult Directed Task assessment sheet

Objective(s): *Find the difference between a positive and a negative*   Date: *07/01/2008*   Adult: *Judith Smith*

integer (Level 4), or two negative integers (Level 5) in context   NC Level: *4 and 5*   Class: *6P*

Child's name	Success criteria				Other observations	Objective(s) achieved	Future action
	Read and interpret negative numbers in context	Find the difference between a positive and a negative integer	Find the difference between two negative integers				
Justin Roach							
Katie Molyneaux							
Sam Perhar							
Alyson Collins							

● Use **Record-keeping format 1** to record individual children's performance during the task, commenting upon particular strengths and weaknesses, how competent you feel the children are with this objective(s) and any future action you may consider appropriate.

Section 2: Pupil Self assessments

Purpose

● To provide children with the opportunity to undertake some form of self assessment at the end of a unit.

List of assessment criteria

Icons for child's assessment

Thought box for child to record something special you want them to think about

Year 6 Self assessment Unit C3

Self assessment Unit C3

Name _____ Date _____

● I can collect and present data in a variety of ways and use my results to solve problems   ☺ ☺ ☹

● I can represent data in a variety of ways and answer questions about the data including interpreting pie charts   ☺ ☺ ☹

● I can use the different averages to solve problems   ☺ ☺ ☹

● I can use the language of chance and likelihood   ☺ ☺ ☹

● I can convert measures between units including decimals   ☺ ☺ ☹

● I can read and interpret scales   ☺ ☺ ☹

● I can compare readings from different scales   ☺ ☺ ☹

● I can solve problems involving more than one step   ☺ ☺ ☹

When to use this feature

- At the end of each unit.

How to use this feature

- Distribute the relevant Pupil Self assessment sheet at the end of the unit.
- The empty thought box at the bottom of the sheet is designed to be used by the children to record anything special that you might like them to think about, e.g.
  - anything they feel they need more practice on
  - what they think they should or could learn next
  - any special equipment that they used to help them during the unit
  - anything they particularly liked or disliked that they did during the unit.
- Ask the children to complete the sheet independently.
- After the children have completed the sheet, as a class, discuss specific objectives, asking individual children to comment on what they have written.

Section 3: Tests

Purposes

- To provide an indication of how individual children, and the class as a whole, have performed during a term, and to inform future planning.
- To inform teacher assessment when assigning an overall National Curriculum Level in mathematics.
- To inform teacher assessment when assigning a National Curriculum Level for each Attainment Target in mathematics.

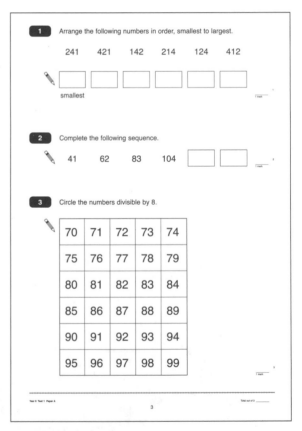

When to use this feature
- At the end of each term.

How to use this feature
- Distribute the relevant Test towards the end of the term. Ensure that all children have the necessary resources.
- Ask the Mental mathematics test questions. This is designed to take approximately 20 minutes.
- Children work independently to complete Papers A and B. Each paper should take approximately 45 minutes.
- Mark the papers and record individual children's results on their paper. You may wish to use **Record-keeping formats 2**, **3** or **4** to analyse the performance of individual children and particular test questions.

Record-keeping format 2 Test 1 grid for test analysis (Mental mathematics test)

Name	Addition: multiples of 10	Multiplication tables	Subtraction: 100 less	2-D shape	Rounding decimals	Calculating positive and negative numbers	Addition: multiples of 10	Addition	Addition: decimals	Fractions/decimals	3-D solids	Length: multiplication	Subtraction: decimals	Addition: fractions	2-D shape/perimeter	Multiplication: money	Multiplication: multiples of 10	Addition/Multiplication	Subtraction	Multiplication/addition: money	Mental mathematics test score (out of 20)
AT	2	2	2	3	2	2	2	2	2	2	3	2 & 3	2	2	3	1 & 2	2	2	2	1 & 2	
Question	1	2	3	4	5	6	7	8	9	10	11	12	13	14	15	16	17	18	19	20	
Mark	1	1	1	1	1	1	1	1	1	1	1	1	1	1	1	1	1	1	1	1	
1. Justin Roach	/	0	/	/	0	/	0	/	0	/	/	0	/	/	0	/	0	/	/	/	19
2. Katie Molyneaux	/	0	0	/	0	/	0	/	0	0	/	/	0	/	/	0	0	0	0	0	8
3. Sam Perhar	/	/	/	/	/	0	/	/	/	/	/	/	0	/	/	/	/	/	/	0	17
4. Alysen Collins																					18
5.																					
6.																					
7.																					
8.																					
9.																					
10.																					
11.																					
12.																					
13.																					
14.																					
15.																					
16.																					
17.																					
18.																					
19.																					
20.																					
21.																					
22.																					
23.																					
24.																					
25.																					
26.																					
27.																					
28.																					
29.																					
30.																					
Number correct																					
Number incorrect or omitted																					
Percentage correct																					
Percentage incorrect or omitted																					

- Using the National Curriculum Level Indicator, assign a Level for the Test.

National Curriculum Level Indicator			
Below Level 3	Level 3	Level 4	Level 5
0–20	21–40	41–80	81–100

* These Tests must be seen only as a guide to help gaining an overall best fit in mathematics.

● Use your professional judgement of each child's overall performance during the term in each of the National Curriculum Attainment Targets. Take into account the following:
  – performance in the Test
  – observations made during Adult Directed Tasks
  – mastery of objectives from the PNS *Renewed Framework for Mathematics* (2006)
  – performance in whole class discussions
  – participation in group work
  – work presented in exercise books
  – any other written evidence.

You may wish to use the following record-keeping formats to assign a Level for each of the National Curriculum Attainment Targets:

Record-keeping format	National Curriculum Attainment Target
Record-keeping format 5	Attainment Target 1 – Using and applying mathematics
Record-keeping format 6	Attainment Target 2 – Number and algebra
Record-keeping format 7	Attainment Target 3 – Shape, space and measures
Record-keeping format 8	Attainment Target 4 – Handling data

Once you have decided which Level best fits a particular child you may wish to identify how competent a child is at that Level by using the following key:

C	Becoming competent at this Level (Achieving up to $\frac{1}{3}$ of the Level Descriptors)	Lower
B	Competent at this Level (Achieving between $\frac{1}{3}$ and $\frac{2}{3}$ of the Level Descriptors)	Secure
A	Very competent at this Level (Achieving more than $\frac{2}{3}$ of the Level Descriptors)	Upper

Record-keeping format 6 Attainment Target 2 – Number and algebra

Level 2

Numbers and the number system	Calculations	Solving numerical problems
• Count sets of objects reliably • Understand place value (HTU) • Order numbers up to 100 • Recognise sequences of numbers • Recognise odd and even numbers	• Recall addition and subtraction number facts to 10 • Understand that subtraction is the inverse of addition	• Use appropriate operation • Use mental strategies to solve problems involving money and measures

Level 3

Numbers and the number system	Calculations	Solving numerical problems
• Understand place value (ThHTU) • Begin to use decimal notation • Recognise negative numbers • Use simple fractions that are several parts of a whole • Recognise when two fractions are equivalent	• Make approximations • Recall addition and subtraction number facts to 20 • Add and subtract two two-digit numbers mentally • Add and subtract three two-digit numbers using written methods • Recall 2, 3, 4, 5, 10 multiplication tables • Recall division facts corresponding to the 2, 3, 4, 5, 10 multiplication tables	• Solve word problems involving larger numbers • Solve word problems involving multiplication • Solve word problems involving division, including those with a remainder

Katie Molyneaux B

Level 4

Numbers and the number system	Calculations	Solving numerical problems
• Multiply and divide whole numbers by 10 or 100 • Add and subtract decimals to two places • Order decimals to three places • Recognise approximate proportions of a whole • Use simple fractions and percentages to describe proportions of a whole • Recognise and describe number patterns • Recognise and describe a multiple, factor and square	• Recall multiplication facts up to 10 × 10 • Recall division facts corresponding to the multiplication facts up to 10 × 10 • Use efficient written methods for addition and subtraction • Use efficient written methods for short multiplication and division	• Use a range of mental calculation strategies for the four operations • Check the reasonableness of an answer • Begin to use simple formulae expressed in words • Use and interpret co-ordinates in the first quadrant

Sam Perhar A
Justin Roach B

Level 5

Numbers and the number system	Calculations	Solving numerical problems
• Multiply and divide whole numbers and decimals by 10, 100 and 1000 • Order negative numbers • Add and subtract negative numbers • Reduce a fraction to its simplest form • Solve simple problems involving ratio and proportion • Calculate fractional or percentage parts of quantities and measurements	• Use brackets appropriately • Use efficient written methods for addition and subtraction up to 10 000 • Use efficient written methods for long multiplication and division • Use all four operations with decimals to two places	• Check solutions by applying inverse operations • Check solutions by estimating using approximations • Construct and express in symbolic form simple formulae involving one or two operations • Use and interpret co-ordinates in all four quadrants

Alyson Collins C

General comments

Year 6 National Expectations
Start-of-year: Level 3a (4c)
End-of-year: Level 4b

● You may also wish to record children's attainment in each of the Key objectives (also referred to as end-of-year expectations) using either of the following record-keeping formats:

Record-keeping format 9: Class record of the end-of-year expectations

Record-keeping format 9 Class record of the end-of-year expectations

Class: *6 P*
Date: *16/05/2008*

Year 6
End-of-year expectations

	Alyson Collins	Katie Molyneaux	Sam Perhar	Justin Roach																	
Counting and understanding number (AT2) Express one quantity as a percentage of another; find equivalent percentages, decimals and fractions (Level 4)																					
Knowing and using number facts (AT2) Use knowledge of place value and multiplication facts to 10 × 10 to derive related multiplication and division facts involving decimals (Level 4)																					
Calculating (AT2) Use efficient written methods to add and subtract integers and decimals, to multiply and divide integers and decimals by a one-digit integer, and to multiply two- and three-digit integers by a two-digit integer (Level 4)																					
Understanding shape (AT3) Visualise and draw on grids of different types where a shape will be after reflection, after translations, or after rotation through 90° or 180° about its centre or one of its vertices (Level 4)																					
Measuring (AT3) Select and use standard metric units of measure and convert between units using decimals to two places (Level 4)																					
Handling data (AT4) Solve problems by collecting, selecting, processing, presenting and interpreting data, using ICT where appropriate; draw conclusions and identify further questions to ask (Level 4)																					

NOTES: **Using and applying mathematics (AT1)** is incorporated throughout
End-of-year National Expectations: Level 4b

Record-keeping format 10: Individual child's record of end-of-year expectations

Record-keeping format 10 Individual child's record of the end-of-year expectations   Name: *Justin Roach*

Year 4		Year 5		Year 6		Year 6 progression to Year 7
Counting and understanding number (AT2)						
Use diagrams to identify equivalent fractions; interpret mixed numbers and position them on a number line (Level 3)	*A*	Explain what each digit represents in whole numbers and decimals with up to two places, and partition, round and order these numbers (Level 3)	*A*	Express one quantity as a percentage of another; find equivalent percentages, decimals and fractions (Level 4)	*B*	Use ratio notation, reduce a ratio to its simplest form and divide a quantity into two parts in a given ratio; solve simple problems involving ratio and direct proportion (Level 5)
Knowing and using number facts (AT2)						
Derive and recall multiplication facts up to 10 × 10, the corresponding division facts and multiples of numbers to 10 up to the tenth multiple (Level 4)	*B*	Use knowledge of place value and addition and subtraction of two-digit numbers to derive sums and differences, doubles and halves of decimals (Level 4)	*B*	Use knowledge of place value and multiplication facts to 10 × 10 to derive related multiplication and division facts involving decimals (Level 4)	*B*	Make and justify estimates and approximations to calculations (Level 5)
Calculating (AT2)						
Add or subtract mentally pairs of two-digit whole numbers (Level 3)	*A*	Use efficient written methods to add and subtract whole numbers and decimals with up to two places (Level 4)	*B*	Use efficient written methods to add and subtract integers and decimals, to multiply and divide integers and decimals by a one-digit integer, and to multiply two- and three-digit integers by a two-digit integer (Level 4)	*B*	Use bracket keys and the memory of a calculator to carry out calculations with more than one step; use the square-root key (Level 5)
Develop and use written methods to record, support and explain multiplication and division of two-digit numbers by a one-digit number, including division with remainders (Level 3)	*A*					
Understanding shape (AT3)						
Know that angles are measured in degrees and that one whole turn is 360°; compare and order angles less than 180° (Level 3)	*A*	Read and plot co-ordinates in the first quadrant; recognise parallel and perpendicular lines in grids and shapes; use a set-square and ruler to draw shapes with perpendicular or parallel sides (Level 4)	*B*	Visualise and draw on grids of different types where a shape will be after reflection, after translations, or after rotation through 90° or 180° about its centre or one of its vertices (Level 4)	*B*	Know the sum of angles on a straight line, in a triangle and at a point, and recognise vertically opposite angles (Level 5)
Measuring (AT3)						
Choose and use standard metric units and their abbreviations when estimating, measuring and recording length, weight and capacity; know the meaning of kilo, centi and milli and, where appropriate, use decimal notation to record measurements (Level 3)	*A*	Draw and measure lines to the nearest millimetre; measure and calculate the perimeter of regular and irregular polygons; use the formula for the area of a rectangle to calculate its area (Level 4)	*B*	Select and use standard metric units of measure and convert between units using decimals to two places (Level 4)	*B*	Solve problems by measuring, estimating and calculating; measure and calculate using imperial units still in everyday use; know their approximate metric values (Level 5)
Handling data (AT4)						
Answer a question by identifying what data to collect; organise, present, analyse and interpret the data in tables, diagrams, tally charts, pictograms and bar charts, using ICT where appropriate (Level 3)	*A*	Construct frequency tables, pictograms and bar and line graphs to represent the frequencies of events and changes over time (Level 4)	*A*	Solve problems by collecting, selecting, processing, presenting and interpreting data, using ICT where appropriate; draw conclusions and identify further questions to ask (Level 4)	*B*	Understand and use the probability scale from 0 to 1; find and justify probabilities based on equally likely outcomes in simple contexts (Level 5)

NOTES: **Using and applying mathematics (AT1)** is incorporated throughout

	Foundation Stage	Year 1	Year 2	Year 3	Year 4	Year 5	Year 6
End-of-year National Expectations	1b	1a (2c)	2b	2a (3c)	3b	3a (4c)	4b

Task 1
Using and applying mathematics

Objectives NC AT 1 NC Level 4

- Solve multi-step problems, and problems involving fractions, decimals and percentages; choose and use appropriate calculation strategies at each stage, including calculator use
- Tabulate systematically the information in a problem or puzzle; identify and record the steps or calculations needed to solve it, using symbols where appropriate; interpret solutions in the original context and check their accuracy
- Explain reasoning and conclusions, using words, symbols or diagrams as appropriate

Resources

- RCM 1: Word problem cards (enlarged to A3 and cut out)
- pencil and paper (per child)
- calculator – optional (per child)

Task

- Give each child one of the Word problem cards from RCM 1 and a pencil and a piece of paper. See below for guidance as to which card to give to individual children depending on their ability.

	Easy	**Moderate**	**Difficult**
Cards	1–6	7–14	15–20

Success criterion: *Read and understand the problem*

- Ask the children to read the card quietly to themselves.
- In turn, go around the group asking each child to explain their word problem to the rest of the group in their own words. Ask: **Jody, what is your problem about? Tim, what do you have to find out?**

Success criterion: *Correctly identify which operation(s) to use*

- Ask each child to suggest which operation(s) they need to use to work out the answer to the word problem. Ask: **Jody, which operation(s) do you need to use to work out the answer to your problem?**
- Ask each child to explain how they know which operation to use. Ask: **Tim, how do you know you need to add/subtract/multiply/divide? What clues are there in the problem?**

Success criterion: *Carry out the calculation(s) using an appropriate method*

- Ask each child to write down the calculation needed to solve the problem and to work out the answer. Say: **On your sheet of paper I want each of you to write down the calculation(s) needed to solve your problem and then I want you to work out the answer.**
- After sufficient time, ask each child to read out the calculation and answer to their problem.
- Encourage each child to talk about the method they used to obtain their answer. Ask: **Sarah, how did you work that calculation out? Show us your working. Is there another way you could have worked it out?**

Success criterion: *Check answer using an effective method*

- Ask each child to check their answer. Say: **I want each of you now to check your answer**.
- After sufficient time ask each child to explain the method they used for checking their answer.

- Repeat the above for the other word problems.
- Conclude by giving each child a calculation, e.g. (16 + 6) × 5, 17·5 × 120, 32 × 18 − 150, 57·2 ÷ 8, and asking them to make up a word problem that can be solved using the calculation. Say: **Sarah, here is your calculation: 32 × 18 − 150. Tell us a word problem using this calculation.**

- How did you know you needed to add/subtract/multiply/divide? What clues were there in the problem?
- How did you decide which part to do first?
- What are the important things you need to remember when solving word problems?

What to do for those children who achieve *above* expectation
- Use the grid opposite to choose suitable Word problem cards.

What to do for those children who achieve *below* expectation
- Use the grid opposite to choose suitable Word problem cards.

Answers

Question	Problem involving	Number of steps required Operation(s) required	Answer	Question	Problem involving	Number of steps required Operation(s) required	Answer
1.	Measures: Capacity	1 step: multiplication	25 million litres	11.	Measures: Area and Money	2 steps: multiplication and multiplication	£416
2.	Measures: Area	1 step: division	5	12.	Measures: Weight	1 step: multiplication	2 kg 960 g
3.	Measures: Time	2 steps: subtraction, subtraction	1 min 10 sec	13.	Measures: Time	1 step: addition or subtraction	11 hours
4.	Measures: Length	1 step: division	1·88 metres	14.	Money	1 step: addition	£5923
5.	Measures: Capacity	2 steps: division and multiplication	300 ml	15.	Money	2 steps: division and multiplication	20%
6.	Measures: Length	1 step: addition or subtraction	8·3 metres	16.	Money	3 steps: division, multiplication and addition	£76 560
7.	Measures: Time and Money	2 steps: division and multiplication	£3	17.	Measures: Weight	3 steps: division, subtraction and multiplication	1·5 kg
8.	Real life	2 steps: division and division	650	18.	Real life	2 steps: division and multiplication	13 230
9.	Real life	2 steps: addition and subtraction	13 698	19.	Real life	2 steps: subtraction and multiplication	18
10.	Money	3 steps: multiplication, division and multiplication	£76.80	20.	Real life	2 steps: addition and subtraction	4·8

Using and applying mathematics

Objective NC AT 1 NC Level 4
- Suggest, plan and develop lines of enquiry; collect, organise and represent information, interpret results and review methods; identify and answer related questions

See Task 25
Handling data and Using and applying mathematics
Page 61

Task 2
Using and applying mathematics

Objectives   NC AT 1   NC Level 4

- Represent and interpret sequences, patterns and relationships involving numbers and shapes; suggest and test hypotheses; construct and use simple expressions and formulae in words then symbols, e.g. the cost of c pens at 15 pence each is 15c pence
- Tabulate systematically the information in a problem or puzzle; identify and record the steps or calculations needed to solve it, using symbols where appropriate; interpret solutions in the original context and check their accuracy
- Explain reasoning and conclusions, using words, symbols or diagrams as appropriate

Resources

- RCM 2: Puzzles 1   ● RCM 3: Puzzles 2   ● pencil and paper (per child)
- calculator – optional (per child)

Task

- Prior to the task, decide which puzzle/investigation to give individual children from RCM 2 and RCM 3. Alternatively, use puzzles or investigations of your own.

> Success criteria: *Represent and interpret sequences, patterns and relationships*
> *Construct and use simple expressions and formulae in words and symbols*
> *Explain reasoning and conclusions in words, symbols and diagrams*

- Give each child a puzzle/investigation from RCM 2 or RCM 3 and a pencil and a piece of paper.
- Ask each child to read through their puzzle/investigation. Ask: **Louise, what is your puzzle/ investigation about? What do you have to find out? What do you know already that can help you solve this?**
- Briefly discuss the puzzle/investigation with each child.
- Say: **I now want each of you to work on your puzzle/investigation. If you need anything, or are unsure of something just ask me.**
- Allow the children sufficient time to spend on their puzzle/investigation. As the children work through the task, ask specific questions to help individual children with the task as well as to assess children's ability to interpret and complete the task.
- Once each child has completed their task, ask each child to talk about what they did and what they found out.
- Encourage each child to talk about any expressions or formulae they constructed. Ask: **Can you tell me the rule you discovered? Did you write a formula? Is your formula in words or symbols? Does your formula always work? How can you be sure?**
- Finally, ask each child to justify why they worked the way they did. Encourage them to explain their methods of working and recording. Ask: **Why did you...? How else could you have gone about it? What did you find easy/difficult about what you did? If you had to do this puzzle/investigation again, how would you do it differently next time?**

● What do you have to do in this puzzle/investigation?
● Explain to me what you have discovered.
● Write about what you discovered. Can you explain this using symbols instead of words?

What to do for those children who achieve *above* expectation

● Provide more challenging puzzles/investigations.
● Encourage the children to construct expressions and formulae in symbols (Level 5).

What to do for those children who achieve *below* expectation

● Provide extra support as the children work through the puzzle/investigation.
● Choose easier puzzles/investigations.

Task 3
Counting and understanding number

Objective
NC AT 2   NC Levels 4 & 5
- Find the difference between a positive and a negative integer (Level 4), or two negative integers (Level 5), in context

Resources
- RCM 4: Temperatures (enlarged to A3) ● 2 small counters
- −10 to 10 number line (for children achieving *below* expectation)

Task
- Place RCM 4 on the table in front of the children.

  Success criterion: *Read and interpret negative numbers in context*

- Place one of the counters on one of the negative numbers and the other on one of the positive numbers.
- Ask: **Which temperature is the warmer temperature? or ... cooler temperature?**

  Success criterion: *Find the difference between a positive and a negative integer*

- Ask: **What is the difference in temperature?**
- Ask the children to explain their method of working, ask: **How did you work that out?**
- Continue until each child has sufficiently demonstrated their ability to find the difference between a positive and a negative integer.

  Success criterion: *Find the difference between two negative integers*

- Now place the two counters onto two different negative numbers.
- Ask: **Which temperature is the warmer temperature? or ... cooler temperature?**
- Ask: **What is the difference between these two temperatures?**
- Ask the children to explain their method of working, ask: **How did you get the answer?**
- Continue until each child has sufficiently demonstrated their ability to find the difference between two negative integers.

- Tell me two temperatures that lie between 0 °C and −10 °C. Which is the warmer? How can you tell? What is the difference between the warmer temperature and −10 °C?
- At six o'clock this morning it was −2 °C. At one o'clock the temperature was 6 °C. By how many degrees did the temperature rise?

What to do for those children who achieve *above* expectation

- Ask the children to find the sum of a positive and a negative integer (Level 4).
- Ask the children to find the sum of two negative integers (Level 5).

What to do for those children who achieve *below* expectation

- Ask the children to order a set of positive and negative integers (Level 3).
- Provide the children with a −10 to 10 number line to help them work out the difference between a positive and a negative integer and two negative integers.

Task 4
Counting and understanding number

Objective	NC AT 2	NC Level 4

• Use decimal notation for tenths, hundredths and thousandths; partition, round and order decimals with up to three places (and position them on the number line)

Resources

● RCM 5: Decimal numbers (enlarged to A3)   ● 10 small counters
● pencil and paper (per child and adult)

Task

● Place RCM 5 on the table in front of the children and provide each child with a pencil and a piece of paper.
● Point out to the children that this sheet has a range of decimals with one-, two- and three-decimal places.

> Success criterion: *Know what each digit in a decimal with up to three places represents*

● Point to specific decimals on the RCM and ask: **Louis, what is this number? Rebecca, tell me this number.**
● Point to a specific digit in a decimal and ask: **Rav, what does this digit represent? What is the value of the five in this number?**
● Repeat several times for each child.
● Point to a different decimal and ask: **William, point to the digit that shows how many units / tenths / hundredths / thousandths are in this number.**
● Repeat the above until each child has sufficiently demonstrated their ability to partition decimals with up to three places.

> Success criterion: *Order a mixed set of decimals with up to three places*

● Say: **I'm going to place one of these counters on 6 different numbers on the sheet. I want you to look at these 6 numbers and write them down on your sheet in order, starting with the largest.**
● Place a counter on six decimals on the RCM.
● Children write down the numbers on their sheet of paper in order, largest to smallest.
● Once each child has done this, discuss the correct order of the numbers with the children.
● Repeat the above, increasing the set of decimals to order from six to ten decimals if appropriate.
● Occasionally ask the children to order the numbers smallest to largest.
● Continue until each child has sufficiently demonstrated their ability to order a mixed set of decimals with up to three places.

Success criteria: *Round a decimal with up to three places to the nearest whole number*
Round a decimal with two or three places to the nearest tenth

- Point to specific decimals on the RCM and ask: **Lucinda, what is 3·912 rounded to the nearest whole number?**
- Then ask: **Lucinda, what is 3·912 rounded to the nearest tenth?**
- Repeat the above until each child has sufficiently demonstrated their ability to round a decimal with up to three places to the nearest whole number and a decimal with two or three places to the nearest tenth.

- What does the digit 6 represent in the number 6·731? What about the 3? ...7? ...1?
- What if it was a length given in metres? ...a mass given in kilograms? ...a capacity in litres?
- What is 3·912 rounded to the nearest whole number? ...rounded to the nearest tenth?

What to do for those children who achieve *above* expectation

- Ask the children to multiply and divide whole numbers and decimals with up to three places by 10, 100 and 1000 and explain the effect (Level 5).

What to do for those children who achieve *below* expectation

- Ask the children to partition, round and order decimals with up to two places (Level 3).

Task 5
Counting and understanding number

Objective   NC AT 2   NC Levels 4 & 5
- Express a larger whole number as a fraction of a smaller one, e.g. recognise that 8 slices of a 5-slice pizza represents $\frac{8}{5}$ or $1\frac{3}{5}$ pizzas (Level 4); simplify fractions by cancelling common factors (Level 5); order a set of fractions by converting them to fractions with a common denominator (Level 4)

Resources
- RCM 6: Reducing fractions (enlarged to A3)   ● 2 × 0–9 dice (preferably in two different colours)
- pencil and paper (per child)

Task
● Provide each child with a pencil and a piece of paper.

> Success criterion: *Express a larger whole number as a fraction of a smaller one*

● Explain to the children that you are going to roll the two 0–9 dice and that each time you do this, you want them to write down the larger number as a fraction of the smaller number.
● Roll the two dice, e.g. 7 and 2, and ask: **Write down the larger number as a fraction of the smaller number.** ($\frac{7}{2}$)
● If the children have written their answer as an improper fraction, ask them to write it as a mixed number. Ask: **If your fraction is an improper fraction, write it for me as a mixed number.** ($3\frac{1}{2}$)
● Continue until each child has sufficiently demonstrated their ability to express a larger whole number as a fraction of a smaller one.

> Success criterion: *Simplify fractions*

● Show the children RCM 6 and the two 0–9 dice. Explain to the children that one of the dice represents the vertical axis and the other dice represents the horizontal axis.
● Children take turns to roll the dice and find the corresponding fraction on the RCM.
● The child then reduces the fraction to its simplest form by cancelling common factors.
● If the child rolls the dice and lands on a fraction that has already been reduced, they roll the dice again.
● Occasionally ask the children to explain their method of working. Say: **Jerry, how did you reduce this fraction to its simplest form?**
● Continue until each child has sufficiently demonstrated their ability to reduce a fraction to its simplest form by cancelling common factors.

> Success criterion: *Order a set of fractions*

● Tell the children that you are going to say three (or if appropriate, four) different fractions to each child and that they are to write these fractions on their sheet of paper as you say them.
● For each child, say three (or four) fractions that have been reduced to their simplest form, e.g. $\frac{3}{4}, \frac{2}{3}, \frac{1}{2}$. (You may wish to refer to the Answers on the next page.)

- Say: **Now I want you to write these fractions out again, but this time I want you to write them in order, starting with the smallest.**
- Continue until each child has sufficiently demonstrated their ability to order a set of three (or four) fractions.

- What fraction of three is eight?
- What do you look for when reducing a fraction to its simplest form?
- How do you know when you have the simplest form of a fraction?
- Give me a fraction equivalent to $\frac{3}{4}$. If the denominator was 24, what would the numerator be?
- Which of these fractions is the smallest? …the largest? How do you know?

What to do for those children who achieve *above* expectation

- After the children have completed the task, point to certain fractions on the grid, e.g. $\frac{9}{21}, \frac{7}{14}, \frac{10}{18}, \frac{3}{12}, \frac{4}{16}$, and ask the children to suggest equivalent fractions, i.e. $\frac{9}{21} = \frac{3}{7} = \frac{6}{14} = \frac{12}{28} = \frac{90}{210}$.

What to do for those children who achieve *below* expectation

- Do not use the two 0–9 dice, instead point to simple fractions, e.g. $\frac{2}{4}, \frac{6}{8}, \frac{2}{20}, \frac{2}{10}, \frac{4}{8}$, and ask the children to reduce the fraction to its simplest form.

Answers

	0	1	2	3	4	5	6	7	8	9
9	$\frac{1}{3}$	$\frac{5}{9}$	$\frac{3}{4}$	$\frac{2}{3}$	$\frac{1}{2}$	$\frac{3}{11}$	$\frac{1}{2}$	$\frac{4}{9}$	$\frac{2}{3}$	$\frac{1}{11}$
8	$\frac{2}{9}$	$\frac{3}{5}$	$\frac{1}{7}$	$\frac{1}{7}$	$\frac{1}{4}$	$\frac{1}{2}$	$\frac{2}{3}$	$\frac{3}{7}$	$\frac{2}{3}$	$\frac{1}{3}$
7	$\frac{1}{5}$	$\frac{5}{8}$	$\frac{2}{5}$	$\frac{3}{8}$	$\frac{1}{3}$	$\frac{6}{7}$	$\frac{2}{5}$	$\frac{3}{4}$	$\frac{1}{3}$	$\frac{1}{4}$
6	$\frac{7}{8}$	$\frac{4}{7}$	$\frac{1}{2}$	$\frac{1}{6}$	$\frac{3}{5}$	$\frac{1}{7}$	$\frac{2}{3}$	$\frac{3}{7}$	$\frac{1}{2}$	$\frac{5}{6}$
5	$\frac{1}{11}$	$\frac{2}{7}$	$\frac{5}{6}$	$\frac{1}{3}$	$\frac{1}{9}$	$\frac{1}{4}$	$\frac{1}{25}$	$\frac{1}{10}$	$\frac{2}{3}$	$\frac{1}{2}$
4	$\frac{5}{9}$	$\frac{3}{10}$	$\frac{1}{6}$	$\frac{1}{6}$	$\frac{4}{9}$	$\frac{5}{6}$	$\frac{1}{3}$	$\frac{2}{3}$	$\frac{4}{9}$	$\frac{1}{6}$
3	$\frac{8}{9}$	$\frac{1}{4}$	$\frac{7}{10}$	$\frac{1}{20}$	$\frac{2}{9}$	$\frac{1}{2}$	$\frac{1}{2}$	$\frac{5}{6}$	$\frac{3}{4}$	$\frac{5}{7}$
2	$\frac{7}{10}$	$\frac{1}{10}$	$\frac{3}{5}$	$\frac{1}{8}$	$\frac{3}{4}$	$\frac{1}{3}$	$\frac{1}{20}$	$\frac{1}{8}$	$\frac{1}{5}$	$\frac{3}{5}$
1	$\frac{7}{8}$	$\frac{2}{7}$	$\frac{3}{10}$	$\frac{3}{11}$	$\frac{5}{8}$	$\frac{9}{10}$	$\frac{1}{4}$	$\frac{7}{9}$	$\frac{9}{10}$	$\frac{1}{3}$
0	$\frac{1}{4}$	$\frac{3}{10}$	$\frac{4}{5}$	$\frac{3}{8}$	$\frac{5}{8}$	$\frac{2}{7}$	$\frac{1}{10}$	$\frac{3}{7}$	$\frac{7}{9}$	$\frac{1}{2}$

Task 6
Counting and understanding number

Objective   NC AT 2   NC Level 4
• **Express one quantity as a percentage of another, e.g. express £400 as a percentage of £1000; find equivalent percentages, decimals and fractions**

Resources

- large sheet of paper ● marker ● RCM 7: Fractions, decimals and percentages cards
- pencil and paper (per child)

Task

> Success criterion: *Express one quantity as a percentage of another*

- Remind the children that a fraction or a decimal can be changed into a percentage by multiplying it by 100.
- On the large sheet of paper write: 'What percentage of 200 is 10?'
- Discuss the various suggestions offered by the children.
- Say: **First we express 10 as a fraction of 200**. On the sheet of paper, write: $\frac{10}{200}$
- Say: **Then we express this as a percentage.** On the sheet of paper, write: $\frac{10}{200} \times 100\% = 5\%$
- Repeat the above for other questions, e.g. **What percentage of £150 is £5? / What percentage is 40 of 200?**, until each child has sufficiently demonstrated their ability to express one quantity as a percentage of another.

> Success criterion: *Find equivalent percentages, decimals and fractions*

- Provide each child with a pencil and a piece of paper.
- Using the fractions, decimals and percentages cards from RCM 7, place one fraction card in front of each child.
- Say: **I want each of you to look at the fraction card in front of you, and on your sheet of paper, I want you to write down this fraction and its equivalent decimal and percentage.**
- Once the children have done this, discuss various answers with the group, commenting on any errors.
- Repeat the above placing a decimal card in front of each child and asking them to write down the equivalent fraction and percentage.
- Repeat the above placing a percentage card in front of each child and asking them to write down the equivalent fraction and decimal.
- Continue until each child has sufficiently demonstrated their ability to express equivalent fractions, decimals and percentages.

- What is this fraction as a decimal? What about as a percentage?
- What do you have to do to express this decimal as a percentage?
- How would you work out what percentage of 1 litre 60 ml is?

What to do for those children who achieve *above* expectation

● Ask the children to find more complex percentage, decimal and fraction equivalences: $\frac{1}{6}$, $\frac{1}{8}$, 0·34, 0·91, 0·88, 17%, 26% and 99%.

What to do for those children who achieve *below* expectation

● Using the cards from RCM 7, only ask the children to find simple percentage, decimal and fraction equivalences, e.g. $\frac{1}{2}$, $\frac{1}{10}$, $\frac{1}{100}$, 0·4, 0·3, 0·8, 0·25, 20%, 70% and 75%.

● Accept answers where fractions are not expressed in their simplest form.

Answers for RCM 7

Fraction*	Decimal	Percentage
$\frac{1}{2}$	0·5	50%
$\frac{1}{10}$	0·1	10%
$\frac{3}{5}$	0·6	60%
$\frac{9}{10}$	0·9	90%
$\frac{1}{100}$	0·01	1%
$\frac{1}{25}$	0·04	4%

Fraction*	Decimal	Percentage
$\frac{1}{4}$	**0·25**	25%
$\frac{2}{5}$	**0·4**	40%
$\frac{3}{10}$	**0·3**	30%
$\frac{4}{5}$	**0·8**	80%
$\frac{1}{50}$	**0·02**	2%
$\frac{3}{20}$	**0·15**	15%
$\frac{31}{50}$	**0·62**	62%

Fraction*	Decimal	Percentage
$\frac{2}{25}$	0·08	**8%**
$\frac{1}{5}$	0·2	**20%**
$\frac{7}{10}$	0·7	**70%**
$\frac{3}{4}$	0·75	**75%**
$\frac{1}{20}$	0·05	**5%**
$\frac{3}{25}$	0·12	**12%**
$\frac{6}{25}$	0·24	**24%**

* Note: fractions have been reduced to their simplest form

Task 7
Counting and understanding number

| Objective | NC AT 2 | NC Level 4 |

• Solve simple problems involving direct proportion by scaling quantities up or down

Resources
- RCM 8: Ratio and proportion word problem cards (enlarged to A3 and cut out)
- pencil and paper (per child)

Task

> Success criterion: *Solve problems involving direct proportion*

- Give each child one of the Ratio and proportion word problem cards from RCM 8 and a pencil and a piece of paper. See below for guidance as to which card to give to individual children depending on their ability.

	Easy	**Moderate**	**Difficult**
Cards	1–6	7–14	15–20

- Ask the children to read the card quietly to themselves.
- In turn, go around the group asking each child to explain their word problem to the rest of the group in their own words. Ask: **Imran, what is your problem about? Sunita, what do you have to find out?**
- Ask each child to suggest how they will work out the answer to the problem. Ask: **Imran, how do you think you might work out the answer to your problem?**
- Ask each child to work out the answer to the problem. Say: **On your sheet of paper I want each of you to solve your problem.**
- After sufficient time ask each child to say the answer to their problem to the group and to talk about how they worked it out. Ask: **Sarah, what is the answer to your problem? How did you work it out? Show us your working. Is there another way you could have worked it out?**
- Repeat the above for the remaining word problems.
- Continue until each child has sufficiently demonstrated their ability to solve simple problems involving direct proportion.

- How did you work out the answer to that problem?
- How else could you have worked it out?

What to do for those children who achieve *above* expectation
- Ask the children to solve problems involving ratio and proportion (other than direct proportion) (Level 5).
- Use the grid above to choose suitable Ratio and proportion word problem cards.

What to do for those children who achieve *below* expectation

● Use the grid on the previous page to choose suitable Ratio and proportion word problem cards.

Answers

1. 2 hours
2. 60p
3. 200 ml
4. 4
5. £4.20
6. £1.20
7. £2.40
8. 125 g
9. 120 g
10. 250 g
11. 60%
12. £4
13. 9
14. 16
15. a) 50 g
    b) 175 g
16. a) 32
    b) 30
17. a) $\frac{1}{4}$
    b) 5 : 7
18. a) 100 ml lime juice
       500 ml fruit purée
       400 ml soda water
    b) 1 litre 200 ml
19. 1 kg
20. a) £4.50
    b) £3.78

Task 8
Knowing and using number facts

Objective NC AT 2   NC Level 4
- Use knowledge of place value and multiplication facts to 10 × 10 to derive related multiplication and division facts involving decimals, e.g. 0·8 × 7, 4·8 ÷ 6

Resources
- RCM 9: Calculating with decimals (enlarged to A3) ● counter (per child) ● 0–9 die
- pencil and paper – optional (per child)

Task

NOTE: If appropriate, allow the children to use pencil and paper to record their thinking and help them to work out the answers.

● Place RCM 9 on the table in front of the children.

> Success criterion: *Calculate: U·t × U*

● Ask each child to place their counter onto one of the decimals on the sheet.
● Say: **I'm going to roll this die. I then want each of you to multiply the number your counter is on by the die number.**
● Ask individual children: **What is the answer to your calculation?**
● Occasionally ask the children to explain their method of working, ask: **Vijay, how did you work that out?**
● Continue until each child has sufficiently demonstrated their ability to calculate: U·t × U.

> Success criterion: *Calculate U·t × T*

● Repeat the above, however this time tell the children a multiple of 10 to multiply their number by, e.g. 3·7 × 40, 6·3 × 30, 0·5 × 20 or 0·7 × 90.
● Continue until each child has sufficiently demonstrated their ability to calculate: U·t × T.

> Success criterion: *0·t × 0·t*

● Repeat the above, however this time ask each child in turn to place two counters onto two different calculators (i.e. decimals with zero units), and to multiply the two numbers together, e.g. 0·5 × 0·3.
● Continue until each child has sufficiently demonstrated their ability to calculate: 0·t × 0·t.

> Success criterion: *U·t ÷ U*

● Repeat, however this time rolling the die first and choosing decimals from the mobile phones for the children to divide. Be sure that the decimal you choose can be divided equally by the die number, e.g. roll 6, and point to 4·2 (answer 0·7), 2·4 (answer 0·4), 1·8 (answer 0·3) or 4·8 (answer 0·8).
● Continue until each child has sufficiently demonstrated their ability to calculate: U·t ÷ U.

● What do you know already that can help you work out the answer to this calculation?
● How many different multiplication and division facts can you make using what you know about 72? What facts involving decimals can you derive?
● What if you started with 7·2? What about 0·72?
● The answer to a calculation is 0·56. What could the calculation be?

What to do for those children who achieve *above* expectation

● Ask the children to calculate questions such as: TU·t × U, e.g. 83·6 × 7.

What to do for those children who achieve *below* expectation

● Do not ask the children to do the second or third part of the task, i.e. calculate: U·t × T or 0·t × 0·t.

Task 9
Knowing and using number facts

Objective   NC AT 2   NC Level 4
- Use knowledge of multiplication facts to derive quickly squares of numbers to 12 × 12 and the corresponding squares of multiples of 10

Resources
- 1–12 die ● large sheet of paper ● marker

Task

> Success criterion: *Recognise squares up to 12 × 12*

- Roll the die, e.g. 6, and ask the children to multiply the number by itself. Ask: **What is six times six? What is six squared? Multiply six by itself.**
- After each different square has been given, write the calculation onto the large sheet of paper, i.e. 6 × 6 = 36.
- Continue until each child has sufficiently demonstrated their ability to recognise squares up to 12 × 12.

> Success criterion: *Squares of multiples of 10*

- Referring to each of the squares written on the large sheet of paper, point to each one in turn and say to individual children: **If six times six equals 36, what is 60 × 60?**
- Then ask: **How did you work out that answer?**
- Continue until each child has sufficiently demonstrated their ability to recognise squares of multiples of 10.

- What is the answer to 5 times 5? What about 50 × 50?
- What is seven squared? Multiply three by itself?
- How could you use 12 × 12 = 144 to work out 13 × 12?

What to do for those children who achieve *above* expectation
- Ask the children to calculate squares up to 20 × 20, explaining their methods of working.

What to do for those children who achieve *below* expectation
- Only ask the children to recognise squares up to 10 × 10.

Task 10
Knowing and using number facts

Objective	NC AT 2	NC Level 5

* Recognise that prime numbers have only two factors and identify prime numbers less than 100; find the prime factors of two-digit numbers

Resources
● large sheet of paper ● marker

Task

> Success criterion: *Recognise that prime numbers have only two factors*

● Remind the children that a factor is a number that divides exactly into another number, leaving no remainder.

● On the large sheet of paper write the number 24 and ask the children to tell you all the factors of 24.

● Say: **Tell me all the numbers that divide into 24 without a remainder.** (1, 2, 3, 4, 6, 8, 12, 24)

● Write these beside the number 24.

● Remind the children that numbers in a multiplication calculation are factors and that factors come in pairs. Referring to the factors of 24, show that 24 = 1 $\times$ 24, 2 $\times$ 12, 3 $\times$ 8 and 4 $\times$ 6.

● On the large sheet of paper write the number 7 and ask the children to tell you all the factors of seven.

● Repeat the above for the numbers 11 and 13.

● Ask: **What can you tell me about the numbers 7, 11 and 13? What do these three numbers have in common? Who can tell me another number that only has two factors? Who can tell me some more numbers?**

● Ask: **What do we call these numbers?**

> Success criterion: *Identify prime numbers less than 100*

● Ask: **Who can tell me some other prime numbers?** Write these on the paper with the other prime numbers.

● Ask: **Is 30 a prime number? Why not? What about 27? Why not?...29? Why? ...41? Why?**

● Repeat asking questions similar to those above until the children have identified the prime numbers less than 100.

> Success criterion: *Find the prime factors of two-digit numbers*

● On the large sheet of paper write the number 28 and say: **Tell me all the factors of 28.** (1, 2, 4, 7, 14, 28)

● Write these on the sheet of paper beside the number 28.

● Then ask: **Which of these factors are prime numbers?** (2 and 7). Draw a ring around these numbers.

● Repeat for other numbers, e.g. 36 and 42.

- Tell me a prime number less than 100. How do you know it is a prime number? Can you tell me another prime number?
- What do these two numbers have in common?
- How many prime factors has 16? What about 20?
- Can you give me a number with the prime factors 3 and 5? What about 2 and 3?

What to do for those children who achieve *above* expectation

- Ask the children to find the prime factors of three-digit numbers, e.g. 112, 132, 144, 124, 162.

What to do for those children who achieve *below* expectation

- Ask the children to identify all the prime numbers to 20 only (Level 4).

Answer

There are 25 prime numbers less than 100: 2, 3, 5, 7, 11, 13, 17, 19, 23, 29, 31, 37, 41, 43, 47, 53, 59, 61, 67, 71, 73, 79, 83, 89 and 97.

Task 11
Knowing and using number facts

Objective   NC AT 2   NC Level 5
• Use approximations, inverse operations and tests of divisibility to estimate and check results

Resources
● RCM 10: Estimate, calculate and check (per child) ● pencil (per child)

Task
NOTE: Prior to the activity, write from four to six different calculations in the first column of RCM 10. The calculations should include addition, subtraction, multiplication and division, and be appropriate to individual children's ability:
– add a two-digit number to a three-digit number
– subtract a two-digit number from a three-digit number
– multiply a two-digit number by a two-digit number
– multiply a three-digit number by a two-digit number
– divide a three-digit number by a one-digit number.

> Success criteria: *Estimate results using a variety of strategies*
> *Check results using a variety of strategies*

● Ask the children to look at the calculations on their sheet.
● Tell the children that for each calculation you want them to:
  – estimate the answer first, writing their estimate and any working out in the second column
  – work out the answer, showing all their working in the third column
  – check their answer, writing any working out in the fourth column.
● As the children work through the calculations on the sheet ask them questions similar to those below, assessing their ability to use approximations, inverse operations and tests of divisibility to estimate and check results.

● What is the approximate answer to this calculation? How did you make your estimate?
● What is the answer to this calculation? How did you work it out?
● How close is the actual answer to your estimate? So do you think that your answer is right? Why?
● Is your answer correct? How can you be so sure?

What to do for those children who achieve *above* expectation
● Include calculations that involve written calculations using all four operations with decimals to two places.

What to do for those children who achieve *below* expectation
● Ask the children to add and subtract a pair of two-digit numbers and to multiply and divide a two-digit number by a one-digit number.

Task 12
Calculating

Objective   NC AT 2   NC Level 4

• Calculate mentally with integers and decimals: U·t ± U·t, TU × U, TU ÷ U, U·t × U, U·t ÷ U

Resources

● RCM 9: Calculating with decimals (enlarged to A3) ● 2 counters (per child) ● 0–9 die
● RCM 11: Two-digit number cards ● pencil and paper (for those children who achieve *below* expectation)

Task

NOTES: If appropriate, allow the children to use pencil and paper to record their thinking and help them to work out the answers.

This task involves a number of Success criteria. It is advisable to choose no more than three or four criteria at a time.

● Place RCM 9 on the table in front of the children.

> Success criterion: *Calculate mentally U·t + U·t*

● Ask each child in turn to place their counters onto any two of the decimals on the sheet.
● Referring to these two decimals ask questions such as: **What is the total of these two numbers? Add these two decimals together. What is the total of these two decimals?**
● Occasionally ask the children to explain their method of working, ask: **Sunita, how did you work that out?**
● Continue until each child has sufficiently demonstrated their ability to calculate mentally U·t + U·t.

> Success criterion: *Calculate mentally U·t − U·t*

● Repeat above, asking the children to find the difference between the pair of decimals.
● Continue until each child has sufficiently demonstrated their ability to calculate mentally U·t − U·t.

> Success criterion: *Calculate mentally TU × U*

● Place a two-digit number card from RCM 11 in front of each child.
● Say: **I'm going to roll this die. I then want each of you to multiply the number on your card by the die number.**
● Ask individual children: **What is the answer to your calculation?**
● Occasionally ask the children to explain their method of working, ask: **Thomas, how did you get that answer?**
● Continue until each child has sufficiently demonstrated their ability to calculate mentally TU × U.

> **Success criterion:** *Calculate mentally TU ÷ U*

- Repeat the above, asking the children to divide the two-digit number by the die number.
- Continue until each child has sufficiently demonstrated their ability to calculate mentally TU ÷ U.

> **Success criterion:** *Calculate mentally U·t × U*

- Referring to RCM 9, ask each child to place one of their counters onto one of the decimals on the sheet.
- Say: **I'm going to roll this die. I then want each of you to multiply the number your counter is on by the die number.**
- Ask individual children: **What is the answer to your calculation?**
- Occasionally ask the children to explain their method of working, ask: **Omar, how did you work that out?**
- Continue until each child has sufficiently demonstrated their ability to calculate mentally U·t × U.

> **Success criterion:** *Calculate mentally U·t ÷ U*

- Repeat the above, however this time rolling the die first and choosing decimals from the mobile phones for the children to divide. Be sure that the decimal you choose can be divided equally by the die number, e.g. roll 6, and point to 4·2 (answer 0·7), 2·4 (answer 0·4), 1·8 (answer 0·3) or 4·8 (answer 0·8).
- Continue until each child has sufficiently demonstrated their ability to calculate: U·t ÷ U.

- What were the mental calculations that you did to solve this calculation?
- What do you know already that can help you work out the answer to this calculation?

What to do for those children who achieve *above* expectation

- Encourage the children to give rapid responses to the questions posed and to make little or no recordings.

What to do for those children who achieve *below* expectation

- Allow the children to use pencil and paper to record their thinking and help them to work out the answers.

Task 13
Calculating .

Objective NC AT 2   NC Level 4
- **Use efficient written methods to add and subtract integers and decimals, to multiply and divide integers and decimals by a one-digit integer, and to multiply two- and three-digit integers by a two-digit integer**

Resources
- RCM 11: Two-digit number cards
- RCM 12: Three-digit number cards
- RCM 13: Four-digit number cards
- RCM 14: Decimal cards – tenths
- RCM 15: Decimal cards – hundredths
- 0–9 die ● pencil and paper (per child)

Task

NOTE: This task involves a number of Success criteria. It is advisable to choose no more than three or four criteria at a time.

- Prior to the task, shuffle and place each set of number cards into a pile of its own.
- Provide each child with a pencil and a piece of paper.

> Success criterion: *Use efficient written methods to add whole numbers*

- Using the three-digit number cards from RCM 12, choose two cards and place them in front of each child. Say: **Using the pencil and paper I want each of you to add the two numbers together using a written method.**
- Allow the children sufficient time to complete their calculation then ask: **Denis, what is your answer? How did you get your answer? Susan, what's your answer? How did you work it out?**
- Collect and reshuffle the cards.
- Repeat the above, using the two-, three- and four-digit number cards from RCMs 11, 12 and 13, asking the children to use efficient written methods to solve calculations involving:
  ThHTU + HTU
  ThHTU + ThHTU
  addition of more than two numbers
- If appropriate, repeat the above until each child has sufficiently demonstrated their ability to carry out an efficient written method to add two (or more) whole numbers.

> Success criterion: *Use efficient written methods to subtract whole numbers*

- Repeat the above, using the three- and four-digit number cards from RCM 12 and RCM 13, asking the children to use efficient written methods to solve calculations involving:
  HTU − HTU
  ThHTU − HTU
  ThHTU − ThHTU

Success criterion: *Use efficient written methods to add decimals*

● Repeat the above, using the tenths decimal cards from RCM 14 and hundredths decimal cards from RCM 15 to ask the children to use efficient written methods to add two (or more) decimals.

Success criterion: *Use efficient written methods to subtract decimals*

● Repeat the above, using the tenths decimal cards from RCM 14 and hundredths decimal cards from RCM 15 to ask the children to use efficient written methods to subtract decimals.

Success criterion: *Use efficient written methods to multiply whole numbers by a one-digit number*

● Place a three-digit number card from RCM 12 or a four-digit number card from RCM 13 in front of each child.

● Say: **I'm going to roll this die. I then want each of you to multiply the number on your card by the die number.**

● Ask individual children: **What is the answer to your calculation? How did you get that answer?**

● Continue until each child has sufficiently demonstrated their ability to carry out an efficient written method to multiply three- or four-digit whole numbers by a one-digit number.

Success criterion: *Use efficient written methods to divide whole numbers by a one-digit number*

● Repeat the above, using the three- and four-digit number cards from RCM 12 and RCM 13, ask the children to use efficient written methods to divide a three- or four-digit whole number by a one-digit number.

Success criterion: *Use efficient written methods to multiply decimals by a one-digit number*

● Place a tenths decimal card from RCM 14 or a hundredths decimal card from RCM 15 in front of each child.

● Say: **I'm going to roll this die and I want each of you to multiply your card number by the die number.**

● Ask individual children: **What is the answer to your calculation? How did you get that answer?**

● Continue until each child has sufficiently demonstrated their ability to carry out an efficient written method to multiply a decimal with up to two places by a one-digit number.

Success criterion: *Use efficient written methods to divide decimals by a one-digit number*

● Repeat the above, using the tenths decimal cards from RCM 14 and hundredths decimal card from RCM 15 to ask the children to use efficient written methods to divide decimals with up to two places by a one-digit number.

Success criterion: *Use efficient written methods to multiply a pair of two-digit numbers*

● Place a pair of two-digit number cards from RCM 11 in front of each child.

● Say: **Using the pencil and paper I want each of you to multiply the two numbers together using a written method.**

● Allow the children sufficient time to complete their calculation then ask: **Denis, what is your answer? How did you get your answer? Susan, what's your answer? How did you work it out?**

● Continue until each child has sufficiently demonstrated their ability to carry out an efficient written method to multiply a pair of two-digit numbers.

> Success criterion: *Use efficient written methods to multiply a three-digit number by a two-digit number*

● Repeat the above, using a two-digit number card from RCM 11 and a three-digit number card from RCM 12.

● What is the answer to that calculation? How did you work it out?
● Did you make an estimate of what the answer might be first? How did you do it?
● Could you have worked out that calculation using a different method? How?
● Did you check your answer? What did you do?

What to do for those children who achieve *above* expectation

● Ask the children to solve calculations using all four operations with decimals up to two places (Level 5).

What to do for those children who achieve *below* expectation

● Only ask the children to calculate with two- and/or three-digit whole numbers (Level 3).

Task 14
Calculating

Objective   NC AT 2   NC Level 4

- Relate fractions to multiplication and division, e.g. $6 \div 2 = \frac{1}{2}$ of $6 = 6 \times \frac{1}{2}$; express a quotient as a fraction or decimal, e.g. $67 \div 5 = 13{\cdot}4$ or $13\frac{2}{5}$; find fractions and percentages of whole-number quantities, e.g. $\frac{5}{8}$ of 96, 65% of £260

Resources

- RCM 16: Relating fractions to multiplication and division (enlarged to A3)
- RCM 17: Finding fractions of numbers (enlarged to A3)
- RCM 18: Percentages (enlarged to A3)
- 2×0–9 dice ● pencil and paper (per child)
- 2×1–6 dice (for those children who achieve *below* expectation)

Task

NOTE: This task involves a number of quite different Success criteria. It is advisable to choose only one criterion at a time.

> Success criterion: *Relate fractions to multiplication and division*

- Place RCM 16 on the table in front of the children.
- Point out to the children how the RCM contains division calculations, fraction calculations and improper fractions. Explain to the children that they are going to take turns to find matching statements on the RCM.
- Point and say: **On this sheet there are some division calculations, some fraction calculations and some improper fractions. We are going to take turns to find matching statements. It might be a division calculation that matches a fraction calculation; or a fraction calculation that matches an improper fraction; or an improper fraction that matches a division calculation. Got the idea? I'll go first.**
- Point to a division calculation, e.g. $12 \div 3$, and the equivalent fraction calculation, i.e. $\frac{1}{3}$ of 12. Say: **12 divided by 3 is the same as a third of 12.**
- Ask a child to start, pointing to a division calculation, fraction calculation or improper fraction and its equivalent statement.
- Continue around the group assessing each child's ability to relate fractions to multiplication and division.

> Success criterion: *Express a quotient as a fraction or decimal*

- Provide each child with a pencil and a piece of paper.
- Ask each child to do the following:
  – roll two 0–9 dice, e.g. 4 and 3 (If they roll a 0 or 1, they roll the die again.)
  – look at the two digits and make a two-digit number and write it down, e.g. 43
  – roll one of the 0–9 dice again, e.g. 4.
- Once each child has done this, ask them to divide their two-digit number by the one-digit number, i.e. $43 \div 4 =$, and write down the answer with the remainder as a fraction, i.e. $43 \div 4 = 10\frac{3}{4}$.

● Where appropriate, also ask the children to express the remainder as a decimal, e.g. $43 \div 4 = 10.75$

● Continue around the group assessing each child's ability to express a quotient as a fraction or decimal.

Success criterion: *Find fractions of whole numbers*

● Place RCM 17 on the table in front of the children.

● Ask the children to take turns to roll the dice, e.g. 3 and 9. (If they roll a 0 it represents ten not zero and if they roll a double, they roll one of the dice again.)

● Explain to the children that the larger number represents the denominator and the smaller number represents the numerator, i.e. $\frac{3}{9}$.

● Looking at the fraction that the child has made, e.g. $\frac{3}{9}$, point to a number on the RCM that is a multiple of the fraction's denominator, i.e. 18, 27, 36, 45, 54, 63, 72, 81 or 90.

● The child then uses the fraction as an operator to find the fraction of the number indicated, e.g. $\frac{3}{9} \times 81$.

● Occasionally ask the children to explain their method of working. Say: **Leroy, how did you work out that answer?**

● Continue until each child has sufficiently demonstrated their ability to use a fraction as an operator to find fractions of numbers.

Success criterion: *Find percentages of whole numbers*

● Place RCM 18 on the table in front of the children.

● Point to one of the percentage labels at the top of the RCM, e.g. 40%, and one of the stars, e.g. 60.

● Ask a child to work out 40% of 60. Say: **Martha, I want you to work out what 40% of 60 is.**

● Repeat the above several times for each child.

NOTE: The first row of percentage labels is the easiest and the fourth row is the hardest.

● Occasionally ask the children to explain their method of working. Ask: **David, how did you work out what 70% of 130 was?**

● Continue until each child has sufficiently demonstrated their ability to find simple percentages of whole numbers.

● Tell me a division calculation with an answer of 2, 5, 10, 20… How did you work it out?

● What are some fractions of numbers that are equal to 2, 5, 10, 20…? How did you go about working this out? How do they relate to division calculations?

● Tell me the remainder to this calculation as a fraction/decimal.

● How do you work out $\frac{3}{5}$ of 45?

● $\frac{3}{7}$ of a total is 12. What is the total? What other fractions of the total can you calculate?

● What percentages can you easily work out in your head? Why are they easy?

● When finding percentages of numbers or quantities what percentages do you use to help you? How do you use these percentages to work out others?

● Which percentages are easy/difficult to work out? Why?

● How would you find 30% of 250? Is there another way? Are there any others? Which do you find easier? Why?

What to do for those children who achieve *above* expectation

● After the children have found fractions and/or percentages of whole numbers, ask them questions that involve finding fractions and/or percentages of quantities, e.g.

75% of £30

$\frac{2}{3}$ of 20 kg

30% of 4 m

$\frac{3}{5}$ of 2 litres

40% of 10 cm

$\frac{5}{6}$ of 1 hour

● When using RCM 18, choose percentages from the fourth row of percentage labels, i.e. 12%, 18%, 46%, 54%, 62%.

What to do for those children who achieve *below* expectation

● Use two 1–6 dice for the activity designed to assess children's ability to express a quotient as a fraction or decimal.

● Do not use the two 0–9 dice with RCM 17, instead point to a number on the RCM, e.g. 24, and say a unitary fraction, e.g. $\frac{1}{4}$. Ask the children to use the fraction as an operator to find the fraction of the number indicated, i.e. $\frac{1}{4} \times 24$ (Level 3).

● When using RCM 18, choose percentages from the first row of percentage labels, i.e. 1%, 10%, 25%, 50%, 75%.

Task 15
Calculating

Objective   NC AT 2   NC Level 4
• Use a calculator to solve problems involving multi-step calculations

Resources
● RCM 10: Estimate, calculate and check (per child) ● pencil (per child) ● calculator (per child)

Task
NOTE: Prior to the activity, write from four to six different calculations in the first column of RCM 10, similar to those below. The calculations should include a combination of addition, subtraction, multiplication and division, and be appropriate to calculate using a calculator.

$$83 \times 17 + 206 \qquad 4628 \times 7 - 659$$
$$(17 \times 9) + (32 \times 43) \qquad (64 \times 21) + (53 - 38)$$
$$(46 \times 43) + (203 \div 7) \qquad (423 \div 9) \times (405 \div 27)$$
$$(74 \times 42) - (95 + 68) \qquad (64 \times 72) + (1168 \div 8)$$
$$(95 + 62) \times (87 + 22) \qquad (84 \div 4) \times (275 - 86)$$

> Success criteria: *Estimate results using a variety of strategies*
> *Use a calculator effectively*
> *Check results using a variety of strategies*

● Provide each child with a copy of RCM 10 containing from four to six different calculations, a pencil and a calculator.
● Ask the children to look at the calculations on their sheet.
● Tell the children that for each calculation you want them to:
  – estimate the answer first, writing their estimate and any working out in the second column
  – work out the answer using a calculator in the third column, writing down the keys pressed in order to get the answer
  – check their answer, writing any working out in the fourth column.
● As the children work through the calculations on the sheet ask them questions similar to those below, assessing their ability to use a calculator effectively.

● Which keys would you press on a calculator to work out: 24 + 6 × 17?
● What are you going to key into your calculator?
● Look at this calculation. What order are you going to key in the numbers and operations in this calculation?
● Nicola has £50. She buys 3 flowerpots at £12.75 each and a spade at £9.65. How much money does she have left? Show me how you used your calculator to find the answer.
● My calculator shows 4.5. What might the question have been? What if it was about weight/length?

What to do for those children who achieve *above* expectation

● Ask the children to solve calculations involving percentages and/or decimals (Level 5).

What to do for those children who achieve *below* expectation

● Use smaller numbers in the calculations to make estimating and mental approximations easier.

Answers

$83 \times 17 + 206 = 1617$

$(17 \times 9) + (32 \times 43) = 1529$

$(46 \times 43) + (203 \div 7) = 2007$

$(74 \times 42) - (95 + 68) = 2945$

$(95 + 62) \times (87 + 22) = 17\,113$

$4628 \times 7 - 659 = 31\,737$

$(64 \times 21) + (53 - 38) = 1359$

$(423 \div 9) \times (405 \div 27) = 705$

$(64 \times 72) + (1168 \div 8) = 4754$

$(84 \div 4) \times (275 - 86) = 3969$

Task 16
Understanding shape

Objective NC AT 3 NC Level 4
- Describe, identify and visualise parallel and perpendicular edges or faces and use these properties to classify 2-D shapes and 3-D solids

Resources

- RCM 19: 2-D shapes (per child) • red and blue coloured pencil (per child) • ruler (per child)
- set of 3-D solid shapes

Task

- Provide each child with a copy of RCM 19, a red and a blue coloured pencil and a ruler.

  > Success criterion: *Recognise parallel sides in 2-D shapes*

- Ask the children to use their red pencil and a ruler to draw over the dotted lines in each shape to show a pair of parallel sides.
- Say: **Look at all the shapes on the sheet. I want you to look for a pair of parallel sides in each shape. Using your red pencil and a ruler I want you to draw over two lines that are parallel to each other. If you cannot find a pair of parallel sides in any of the shapes, then write a red zero inside that shape.**
- Next ask the children to identify how many pairs of parallel sides each shape has.
- Say: **Look at all the shapes. Now I want you to use your red pencil to write inside each shape how many pairs of parallel sides it has.**
- Continue until each child has sufficiently demonstrated their ability to recognise parallel sides in 2-D shapes.

  > Success criterion: *Recognise perpendicular sides in 2-D shapes*

- Now ask the children to use their blue pencil and a ruler to draw over the dotted sides in each shape to show a pair of perpendicular sides.
- Say: **Look at all the shapes on the sheet again. This time I want you to use the blue pencil and a ruler to draw over two lines on each shape that are perpendicular to each other. If you cannot find a pair of perpendicular sides in any of the shapes, then write a blue zero inside that shape.**
- Next ask the children to identify how many pairs of perpendicular sides each shape has.
- Say: **Look at all the shapes. Now I want you to use your blue pencil to write inside each shape how many pairs of perpendicular sides it has.**
- Continue until each child has sufficiently demonstrated their ability to recognise perpendicular sides in 2-D shapes.

Success criterion: *Recognise parallel and perpendicular edges and faces in 3-D solids*

- Place the set of 3-D solids on the table in front of the children.
- Ask children questions similar to the following: **Look at this cube. How many edges/faces are parallel to this one? How many edges/faces are perpendicular to this one? How many vertices does a cuboid have? How many edges/faces? How many pairs of parallel edges/faces has a cuboid? ...perpendicular edges/faces...?**
- Continue until each child has sufficiently demonstrated their ability to recognise parallel and perpendicular edges/faces in 3-D solids.

- How can you check to make sure two sides/edges/faces are parallel/perpendicular?
- What do you know about the properties of a rectangle?
- What is the same/different about a square and a rectangle? What about this rhombus and kite?

What to do for those children who achieve *above* expectation

- Using their red and blue pencils, ask the children to mark all the pairs of parallel sides and perpendicular sides in each of the 2-D shapes.

What to do for those children who achieve *below* expectation

- Do not ask the children to identify all the pairs of parallel sides and perpendicular sides in each 2-D shape.

Answers

Bold number – Red answer
(Number of pairs of parallel sides)

Italic number – Blue answer
(Number of pairs of perpendicular sides)

Accept any of the marked pairs of parallel and perpendicular sides.

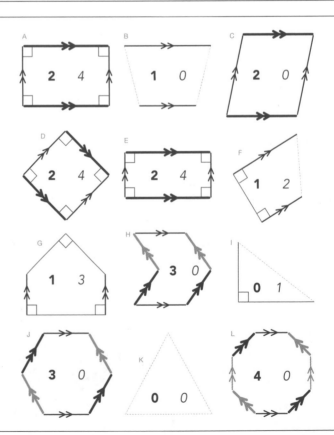

Task 17
Understanding shape

Objective   NC AT 3   NC Level 4
• Make and draw shapes with increasing accuracy and apply knowledge of their properties

Resources
● squared or dot paper (per child) ● pencil (per child) ● ruler (per child) ● set square (per child)
● protractor (per child) ● pair of compasses (per child) ● scissors (per child)

Task
● Provide each child with the resources listed above.

> Success criterion: *Draw polygons with accuracy*

● Ask each child to draw a polygon. You may wish to:
– tell the children the name of the shape, e.g. draw a pentagon that has three right angles
– describe a shape and ask the children to draw it, e.g. a four-sided shape, with opposite sides equal
– give specific measurements, e.g. draw a rhombus with sides of 5 cm.
● Continue until each child has sufficiently demonstrated their ability to draw polygons.

> Success criterion: *Draw a circle*

● Ask each child to draw a circle using a pair of compasses.

> Success criterion: *Draw a net and construct a 3-D solid*

● Ask each child to draw the net of a cube with edges of 6 cm, and then construct the solid.

● How did you draw that shape?
● Describe to me what you would use to draw a regular hexagon? How would you go about drawing it?

What to do for those children who achieve *above* expectation
● Ask the children to make and draw shapes of greater complexity, e.g. regular octagon, net of a tetrahedron.

What to do for those children who achieve *below* expectation
● Ask the children to draw simple 2-D shapes, e.g. square, rectangle and right-angled triangle.

Task 18
Understanding shape

Objective   NC AT 3   NC Level 4
- Visualise and draw on grids of different types where a shape will be after reflection, after translations, or after rotation through 90° or 180° about its centre or one of its vertices

Resources
- RCM 20: Reflect, translate and rotate (per child) ● pencil (per child) ● ruler (per child)
- coloured pencil – optional (per child)

Task
● Provide each child with a copy of RCM 20, a pencil and a ruler.

> Success criterion: *Visualise and draw a shape after a reflection*

● Referring to Grid 1 on RCM 20, draw children's attention to the vertical and diagonal lines of symmetry.
● Ask the children to draw the reflection of one of the following:
– Shape A along the vertical line of symmetry – easier
– Shape B along the diagonal line of symmetry – harder
● If appropriate, ask the children to draw the reflection of the other shape.

> Success criterion: *Visualise and draw a shape after a translation*

● Referring to Grid 2 on RCM 20, ask the children to translate one of the following:
– Shape C four squares to the right – easier
– Shape D five squares to the left and one square up – harder
● If appropriate, ask the children to translate the other shape.

> Success criterion: *Visualise and draw a shape after a rotation through 180° about one of its vertices*

● Referring to Grid 3 on RCM 20, ask the children to rotate Shape E 180° about Point A.

> Success criterion: *Visualise and draw a shape after a rotation through 90° about its centre*

● Referring to Grid 4 on RCM 20, ask the children to rotate Shape F 90° about Point A.

 ● Explain to me how this shape has been reflected. …translated. …rotated.

What to do for those children who achieve *above* expectation

● Ask the children to complete the two harder reflections and translations and to rotate Shape F 90° about its centre (Grid 4).

● Ask the children to complete all six reflections, translations and rotations.

What to do for those children who achieve *below* expectation

● Ask the children to complete the two easier reflections and translations and to rotate Shape E 180° about one of its vertices (Grid 3).

Answers

Grid 1

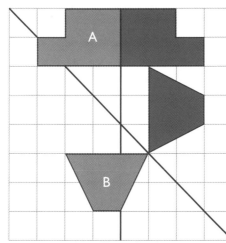

Grid 2

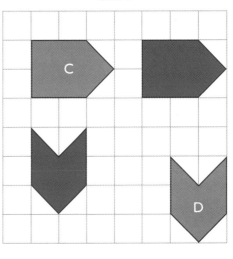

Grid 3

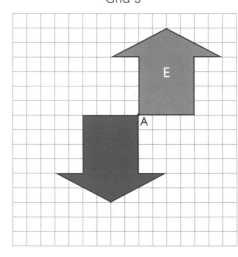

Grid 4

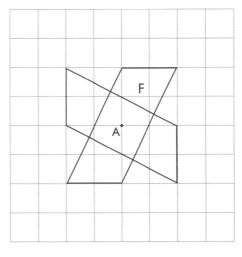

Task 19
Understanding shape

Objective   **NC AT 2 & 3***   **NC Level 4**
- Use co-ordinates in the first quadrant to draw, locate and complete shapes that meet given properties

Resources

- RCM 21: First quadrant co-ordinates (per child) ● pencil (per child) ● ruler (per child)

Task

● Provide each child with a copy of RCM 21, a pencil and a ruler.

> Success criterion: *Use co-ordinates in the first quadrant to draw and locate points*

● Say: **You need to listen carefully. I am going to give you some instructions to follow. I will only say the instructions once, so you must listen carefully the first time. Do only the things you are told to do, and do nothing else. There are 8 instructions. Do each task immediately after the instructions have been given. Ready?**

● Say:
  1. Draw a ring around the cross at co-ordinates (3, 9)
  2. Draw a ring around the cross at co-ordinates (10, 11)
  3. Draw a cross at the co-ordinates (7, 4)
  4. Draw a cross at the co-ordinates (12, 1)
  5. Draw a ring around the cross at co-ordinates (4, 15)
  6. Draw a cross at the co-ordinates (9, 10)
  7. Draw a ring around the cross at co-ordinates (13, 4)
  8. Draw a cross at the co-ordinates (15, 13)

> Success criterion: *Use co-ordinates in the first quadrant to complete shapes*

● Draw children's attention to the lines drawn on the RCM. Explain to the children how these lines make up the sides of four different quadrilaterals.

● Ask the children to use their ruler to complete each of the four quadrilaterals.

● Then ask the children to write down the four sets of co-ordinates for each of the four shapes that identify the shapes' vertices.

● If necessary, choose one of the quadrilaterals and, with the children's help, identify the co-ordinates for one or two of the vertices.

● As the children complete the task, monitor how successful they are at reading and writing co-ordinates in the first quadrant.

*Although this objective appears under AT2 – Number and algebra in the NC, in
the PNS *Framework* it appears under the 'Understanding shape' strand.

● Which number comes first in a pair of co-ordinates, the x-axis or the y-axis?
● How did you find that point on the grid?
  ● A square has vertices at (0, 0), (5, 0) and (5, 5). What are the co-ordinates of the fourth vertex?

What to do for those children who achieve *above* expectation

● Ask the children to read and interpret co-ordinates in all four quadrants (Level 5).

What to do for those children who achieve *below* expectation

● Do not ask the children to do the second part of the task.

Answers

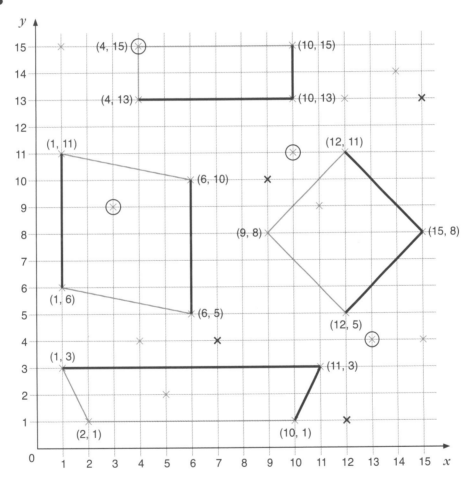

Task 20
Understanding shape

> **Objective**   NC AT 3   NC Levels 4 & 5
> • Estimate angles (Level 4), and use a protractor to measure and draw them, on their own and in shapes
>   (Levels 4 & 5); calculate angles in a triangle or around a point (Level 5)

Resources
- RCM 22: Measuring angle cards (enlarged to A3 and cut out)
- RCM 23: Angles in a triangle and around a point cards (enlarged to A3 and cut out)
- pencil and paper (per child) • ruler (per child) • protractor (per child)

Task
- Provide each child with a pencil and a piece of paper, a ruler and a protractor.

> Success criterion: *Estimate angle*

- Place one of the cards from RCM 22 in front of each child. See below for guidance as to which card to give to individual children depending on their ability.

	Easy (angles to the nearest 10°)	**Moderate** (angles to the nearest 5°)	**Difficult** (angles to the nearest 1°)
Cards	1, 2, 3, 4	5, 6, 7, 8	9, 10, 11, 12

- Ask: **I want each of you to look at the angle drawn on the card in front of you and to tell me whether it is an acute, obtuse or right angle.**
- Once each child has done this, say: **Now I want each of you to estimate the size of the angle on the card.**

> Success criterion: *Use a protractor to measure angles*

- Now ask each child to use their protractor to measure the size of the angle.
- Repeat the above until each child has sufficiently demonstrated their ability to estimate and measure angles.

> Success criterion: *Use a protractor to draw angles*

- Ask individual children to use their ruler and protractor to draw one of the following angles. See below for guidance as to which angle to ask individual children to draw depending on their ability.

Easy (angles to the nearest 10°)	**Moderate** (angles to the nearest 5°)	**Difficult** (angles to the nearest 1°)
50°, 80°, 100°, 120°	25°, 55°, 115°, 135°	8°, 21°, 163°, 114°

- Continue until each child has sufficiently demonstrated their ability to draw angles.

Success criterion: *Calculate angles in a triangle*

● Using only cards 1 to 6 from RCM 23, place one card in front of each child. See below for guidance as to which card to give to individual children depending on their ability.

	Easy (angles to the nearest 10°)	**Moderate** (angles to the nearest 5°)	**Difficult** (angles to the nearest 1°)
Cards	1, 2	3, 4	5, 6

● Say: **Look at the card in front of you. I want each of you to calculate the size of the missing angle.**

● Once the children have done this, ask: **How did you calculate the size of the missing angle? How did you know it was ... degrees?**

Success criterion: *Calculate angles around a point*

● Using only cards 7 to 12 from RCM 23, place one card in front of each child. See below for guidance as to which card to give to individual children depending on their ability.

	Easy (calculate reflex angles)	**Difficult** (calculate angles formed by two straight lines)
Cards	7, 8, 9	10, 11, 12

● Say: **Look at the card in front of you. I want each of you to calculate the size of the missing angle(s).**

● Once the children have done this, ask: **How did you calculate the size of the missing angle(s)? How did you know that this angle was ... degrees? How did you calculate the size of this angle?**

● What things do you have to remember when you use a protractor to measure or draw angles?
● What is the angle between the hands of a clock at four o'clock? Explain how you know.
● Estimate the size of each of these angles. Now measure them to the nearest degree. How close were you?
● Draw me an angle of 47°.
● How many degrees are there in a triangle? ...around a point?
● When two straight lines intersect, what can you tell me about the opposite angles?

What to do for those children who achieve *above* expectation

● Give the children cards 9–12 from RCM 22 to measure angles to the nearest 1° (Level 5).
● Draw several shapes on a piece of paper and ask the children to measure the interior angles of each shape, e.g.

● Ask the children to draw angles to the nearest 1° (Level 5).
● Give the children cards 5, 6, 10, 11 and 12 from RCM 23 to calculate angles in a triangle and around a point.

What to do for those children who achieve *below* expectation

- Give the children cards 1–4 from RCM 22 to measure angles to the nearest 10°.
- Give the children cards 5–8 from RCM 22 to measure angles to the nearest 5°.
- Ask the children to draw angles to the nearest 10° or 5°.
- Give the children cards 1, 2, 3, 4, 7, 8 and 9 from RCM 23 to calculate angles in a triangle and around a point.

Answers

RCM 22: Measuring angle cards

1. 70°	**2.** 30°	**3.** 140°
4. 110°	**5.** 45°	**6.** 75°
7. 125°	**7.** 165°	**8.** 83°
10. 98°	**11.** 172°	**12.** 54°

RCM 23: Angles in a triangle and around a point cards

1. $a = 110°$	**2.** $b = 20°$	**3.** $c = 40°$
4. $d = 55°$	**5.** $e = 51°$	**6.** $f = 64°$
7. $g = 324°$	**8.** $h = 288°$	**9.** $i = 306°$
10. $k = 30°$	**11.** $n = 110°$	**12.** $s = 55°$
$m = 150°$	$p = 70°$	$t = 125°$
	$r = 110°$	$u = 55°$

Task 21
Measuring

Objective   NC AT 3   NC Level 4
- **Select and use standard metric units of measure and convert between units using decimals to two places, e.g. change 2·75 litres to 2750 m/, or vice versa**

Resources
- 3 large sheets of paper ● marker

Task

> Success criterion: *Convert units using decimals to two places – Length*

● Prior to the task divide a large sheet of paper in twelfths and write the following (or similar) measures at the top of each section:

3·42 m	140 cm	12·68 km	24·19 cm
71 mm	8200 m	2·46 m	3400 cm
42·6 km	53·6 cm	74·1 m	704 mm

● Briefly revise children's understanding of multiplying and dividing by 10, 100 and 1000. Ask questions such as: **What is 54 multiplied by 100? What is 340 divided by 10? What is one tenth of 62? What is the product of 18 and 1000? Divide 81 by 1000.** Include questions with both whole number and decimal answers.

● Also, briefly remind the children of the following measurement equivalences:
  1 m = 100 cm = 1000 mm
  1 km = 1000 m

● Point to one of the measurements on the sheet, e.g. 12·68 km. Say: **I want to find out how many metres there are in 12·68 km.** Ask: **What do I need to do to kilometres to get metres? So, how many metres are there in 12·68 km?**

● Write the answer in the same section of the sheet.

● Repeat the above, pointing to other measurements on the sheet until the children have sufficiently demonstrated their ability to convert between different units of length.

Success criterion: *Convert units using decimals to two places – Mass*

● Prior to the task divide a large sheet of paper in twelfths and write the following (or similar) measures at the top of each section.

750 g	542 g	11·8 kg	24·19 kg
1530 g	6700 g	3·52 kg	7·4 kg
6·92 kg	2870 g	800 g	8·35 kg

● Remind the children that 1 kg = 1000 g.
● Repeat the above.

Success criterion: *Convert units using decimals to two places – Capacity*

● Prior to the task divide a large sheet of paper in twelfths and write the following (or similar) measures at the top of each section.

2·35 l	6·41 l	40 cl	600 cl
3520 ml	6200 ml	700 ml	7·6 l
19·64 l	250 cl	950 ml	12·02 l

● Remind the children that 1 litre = 100 cl = 1000 ml.
● Repeat the above.

● How do I write 7 metres 5 centimetres as a decimal?
● Tell me an example of something you would measure in kilometres. What about metres / centimetres / millimetres?
● How do I write 12 litres 600 millilitres as a decimal?

What to do for those children who achieve *above* expectation
● Ask children to convert units using decimals to three places.

What to do for those children who achieve *below* expectation
● Ask children to convert units using decimals to one place only.

Task 22
Measuring

Objective   NC AT 3   NC Level 4

- Read and interpret scales on a range of measuring instruments, recognising that the measurement made is approximate and recording results to a required degree of accuracy; compare readings on different scales, e.g. when using different instruments

Resources

- sheets of A4 paper ● ruler (per child and yourself) ● pencil (per child and yourself)
- RCM 24: Reading and interpreting scales – Mass (enlarged to A3 and cut out)
- RCM 25: Reading and interpreting scales – Capacity (enlarged to A3 and cut out)

Task

- Prior to the task, for each child you are going to be doing the task with, draw about three lines of different lengths and orientations, e.g. horizontal, vertical, diagonal, onto a sheet of paper.

  Success criterion: *Read and interpret scales – Length*

- Provide each child with a sheet of paper showing about three lines of different lengths, a ruler and a pencil.
- Ask each child to use their ruler to measure the three lines on the sheet of paper in front them.
  Say: **Look at the lines drawn on your sheet of paper. I want each of you to measure these lines to the nearest millimetre.**
- When each child has done this, ask individual children to draw lines of various lengths, e.g. Say: **Michael, on your sheet of paper, I want you to draw a line 12·5 cm long. Fabio, I want you to draw a line 168 mm long.**
- Repeat the above until each child has sufficiently demonstrated their ability to measure and draw lines to the nearest millimetre.

  Success criterion: *Read and interpret scales – Mass*

- Place one of the cards from RCM 24 in front of each child. See below for guidance as to which card to give to individual children depending on their ability.

	Easy (indicator pointing to labelled division)	**Moderate** (indicator pointing to unlabelled division)	**Difficult** (indicator pointing between two divisions)
Cards	1, 2, 3	4, 5, 6	7, 8, 9, 10, 11, 12

- Ask the children to look at their card and to read and interpret the scales.
- Repeat the above if necessary.
- Then lay all the cards on the table in front of the children.
- Ask questions similar to the following that involve the children comparing and contrasting the weights shown on different cards, e.g. **Which card shows 3·5 kg? Which of these cards shows the lightest weight? …heaviest weight? Which of these cards shows a weight less than 2·5 kg? …more than 10 kg?**

Success criterion: *Read and interpret scales – Capacity*

● Place one of the cards from RCM 25 in front of each child. See below for guidance as to which card to give to individual children depending on their ability.

	Easy (water level at labelled division)	**Moderate** (water level at unlabelled division)	**Difficult** (water level between two divisions)
Cards	1, 2, 3	4, 5, 6	7, 8, 9, 10, 11, 12

● Ask the children to look at their cards and to read and interpret the scales.

● Repeat the above if necessary.

● Then lay all the cards on the table in front of the children.

● Ask questions similar to the following that involve the children comparing and contrasting the capacities shown on different cards, e.g. **Which card shows 2·5 litres? Which of these cards shows the least amount of liquid? …most amount of liquid? Which of these cards shows a capacity less than 3 litres? …more than 2·25 litres?**

● What is the weight shown on this scale? How do you know?

● How much water is there in this container?

● What does this scale show you?

● What do these unlabelled divisions on this scale represent? How do you know?

What to do for those children who achieve *above* expectation

● Using cards 7–12 on RCM 24 and RCM 25, ask the children to read and interpret scales where the indicator, or water level, is between two divisions.

● Ask questions that involve the children calculating the weights and capacities shown in different cards, e.g. point and ask: **What is the total of these two weights? Which card shows a weight 1·5 kg less than this card? Which of these two cards shows the most amount of water? How much more water is there in this container than in this container?**

What to do for those children who achieve *below* expectation

● Using cards 1–6 on RCM 24 and RCM 25, ask the children to read and interpret scales where the indicator is pointing to, or water level is at, either labelled or unlabelled divisions.

Answers

RCM 24: Reading and interpreting scales – Mass

1. 400 g	**2.** 30 kg	**3.** 1·5 kg	**4.** 1 kg	**5.** 150 g	**6.** 14 kg
7. 3·5 kg	**8.** 50 kg	**9.** 0·9 kg	**10.** 900 g	**11.** 15 kg	**12.** 3 kg

RCM 25: Reading and interpreting scales – Capacity

1. 400 ml	**2.** 9 litres	**3.** 200 ml	**4.** 3·5 *l*	**5.** 1 litre	**6.** 350 ml
7. 2·5 *l*	**8.** 7 litres	**9.** 750 ml	**10.** 4·5 *l*	**11.** 350 ml	**12.** 1·4 *l*

Task 23
Measuring

> **Objective**   **NC AT 3**   **NC Level 4**
> • Calculate the perimeter and area of rectilinear shapes; estimate the area of an irregular shape by counting squares

Resources

- RCM 26: Perimeter and area cards 1 (enlarged to A3 and cut out) ● pencil and paper (per child)
- RCM 27: Perimeter and area cards 2 (enlarged to A3 and cut out) (for children achieving *above* and *below* expectation) ● calculator (per child) (for children achieving *above* expectation)

Task

- Prior to the task, enlarge and cut out the cards from RCM 26 and RCM 27. Place each set of cards in a separate pile. The cards from RCM 27 are only needed for children achieving *above* and *below* expectation.
- Provide each child with a pencil and a piece of paper.

> Success criteria: *Understand how to find the perimeter of simple compound shapes*
> *Calculate the perimeter of compound shapes*

- Place one of the cards from RCM 26 in front of each child. See below for guidance as to which card to give to individual children depending on their ability.

	Easy	Moderate	Difficult
Cards	1, 2, 3	4, 5, 6, 7, 8, 9	10, 11, 12

- Ask the children to find the perimeter of the shape, and to show their working on the piece of paper.
- Repeat the above until each child has sufficiently demonstrated their ability to find the perimeter of simple compound shapes.

> Success criteria: *Understand how to find the area of simple compound shapes*
> *Calculate the area of compound shapes*

- Place one of the cards from RCM 26 in front of each child. See the above for guidance as to which card to give to individual children depending on their ability.
- Ask the children to find the area of the shape, and to show their working on the piece of paper.
- Repeat the above until each child has sufficiently demonstrated their ability to find the area of simple compound shapes.

- Why does splitting a shape into rectangles help you find the perimeter/area?
- How do you go about finding all the dimensions of the compound shape to work out the perimeter/area?
- How did you work out the perimeter/area of this shape? Is there another way you could have done it? Could you have split the shapes into rectangles a different way?

What to do for those children who achieve *above* expectation

● Using cards 7–12 on RCM 27, ask the children to find the perimeter and area of the compound shapes. If necessary, allow the children to use a calculator to carry out the calculations. For card 9, the children find the perimeter and the area of the shaded sections (Level 5).

What to do for those children who achieve *below* expectation

● Using cards 1–6 on RCM 27, ask the children to find the perimeter and area of the rectangles.

Answers

RCM 26: Perimeter and area cards 1

Shape	Perimeter	Area
1	76 cm	264 cm²
2	66 cm	222 cm²
3	78 cm	328 cm²
4	56 cm	136 cm²
5	42 cm	68 cm²
6	100 cm	336 cm²
7	92 cm	299 cm²
8	66 cm	182 cm²
9	172 cm	1488 cm²
10	126 cm	346 cm²
11	108 cm	224 cm²
12	144 cm	720 cm²

RCM 27: Perimeter and area cards 2

Shape	Perimeter	Area
1	76 cm	360 cm²
2	60 cm	176 cm²
3	80 cm	384 cm²
4	60 cm	224 cm²
5	44 cm	85 cm²
6	100 cm	600 cm²
7	79 cm	177·76 cm²
8	89·2 cm	328·6 cm²
9	200 cm	1050 cm²
10	121·4 cm	283·9 cm²
11	188 cm	432 cm²
12	62 cm	164 cm²

Task 24
Handling data

Objective   NC AT 4   NC Level 4
• Describe and predict outcomes from data using the language of chance or likelihood

---

Resources
● RCM 28: Probability (enlarged to A3) ● pack of playing cards (with the picture cards removed)
● large sheet of paper ● marker

Task
● Prior to the task, draw the following table on a large sheet of paper. Number the first column 1 to 20.

	Outcome		
	colour	suit	number
1			
2			
3			
4			

Success criterion: *Describe and make predictions using the language of chance or likelihood*

● Show the children RCM 28. Remind the children of the five different probabilities on the sheet.
● Pointing to each of the five different probabilities in turn, ask: **Rebecca, can you tell me something that is certain to happen? Michelle, can you tell me something that has a good chance of happening?**
● Continue to ask children questions that require them to suggest an event for each of the five probabilities on the sheet.
● Then ask children questions that require them to describe the chance or likelihood of an event, e.g. **What is the probability that a new baby will be a boy? What is the probability that the temperature will be above 30 degrees tomorrow? What is the probability that the number rolled on a 1–6 die will be six?**

Success criterion: *Predict outcomes*

● Place the pack of playing cards (with the picture cards removed) on the table in front of the children.
● Say: **This is a pack of 52 cards with the picture cards removed. So there are now 40 cards in the pack.**
● Ask each child to predict the probability of taking a red card from the pack. Ask: **What do you think is the probability of taking a red card from this pack? I want each of you to write your name followed by the letter 'R' written in a circle, in the box on the sheet where you think it should go.**
● Repeat the above, asking the children to predict the expected probability of taking a spade from the pack. (Name followed by the letter 'S' written in a circle.)
● Repeat the above, asking the children to predict the expected probability of taking an even number from the pack. (Name followed by the letter 'E' written in a circle.)

Success criterion: *Investigate the probability of events*

- Say: **We are now going to investigate the probability of choosing a red card, a spade and an even number.**
- Shuffle the cards and turn over the top card. With the children's help record the results in the table.
- Repeat 20 times.
- Then as a group, discuss the experimental probability of choosing a red card, a spade and an even number.
- Compare the results of the experimental probability with the expected probability.

- What is the probability of tossing a coin and it landing heads up?
- Tell me something that has a poor chance of happening today?

What to do for those children who achieve *above* expectation

- On RCM 28, make the probabilities into a probability scale, i.e.

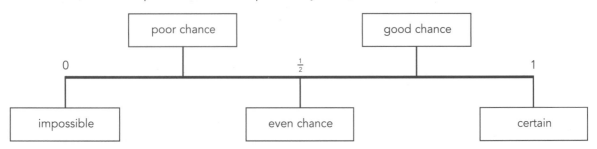

Ask the children to describe and predict outcomes from data using the language of chance or likelihood and the probability scale 0 to 1 (Level 5).

What to do for those children who achieve *below* expectation

- Only ask the children to describe and predict outcomes from data using the language of chance or likelihood.

Task 25

Handling data
Using and applying mathematics

Objectives   NC AT 4 & 1   NC Level 4
- **Solve problems by collecting, selecting, processing, presenting and interpreting data, using ICT where appropriate; draw conclusions and identify further questions to ask**
- Suggest, plan and develop lines of enquiry; collect, organise and represent information, interpret results and review methods; identify and answer related questions

Resources
- RCM 29: Collecting, selecting and organising data (per child)
- squared and/or graph paper (per child) ● pencil (per child) ● ruler (per child)
- ICT data handling package – optional (per child or group)

Task
- Prior to the task, write a question in the box at the top of RCM 29. Choose a topic to investigate that is relevant to your particular circumstances and of interest to the children, e.g. *What is our favourite sport? Is there more football on television than any other sport? How much television do we watch a week? What is the cost of the average house in our local area? What are the five most favourite books in our school? What is our favourite holiday destination? How many different languages are spoken in our school?* Alternatively, you may wish the children to suggest their own line of enquiry.

NOTE: You may want the children to work in pairs for this task.

> Success criteria: *Collect data*
> *Select data*
> *Organise data*
> *Present data*
> *Interpret data*
> *Draw conclusions*
> *Identify further questions to ask*

- Provide each child with a copy of RCM 29, some squared and/or graph paper, a pencil and a ruler.
- Briefly discuss the question with the children.
- Discuss with the children:
  – their initial ideas of what to do
  – how they are going to collect the data
  – how they are going to organise the data.
- Ask the children to collect, organise and present the data.
- When the children have done this, they write about what they found out and what else they could find out about the topic.

● What information will you need to collect to answer this question? How will you collect it?
● What does this graph tell you? What makes the information easy or difficult to interpret?
● Look at this graph, table or chart. Make up three questions that can be answered using the data that is represented.
● What further information could you collect to answer the question more fully?

What to do for those children who achieve *above* expectation

● Ask the children to present their data using an ICT data-handling package.

What to do for those children who achieve *below* expectation

● Ask the children to work in pairs.

Task 26
Handling data

Objective   NC AT 4   NC Level 4

- (Construct and) interpret frequency tables, bar charts with grouped discrete data, and line graphs; interpret pie charts

Resources

- RCM 30: Handling data 1 (enlarged to A3) ● RCM 31: Handling data 2 (enlarged to A3)
- pencil and paper (per child) ● calculator – optional (per child)

Task

> Success criterion: *Extract and interpret information presented in a bar chart or bar line chart*

- Draw children's attention to the bar chart on RCM 30.
- Briefly discuss the bar chart with the children.
- Say: **Look at the bar chart. What does it show?**
- Ask individual children questions similar to the following to assess their ability to extract and interpret information presented in a bar chart.
    - **Between which times of the day does Paul take the most money?**
    - **Approximately what is the difference between the period when Paul takes the most amount of money and the period when he takes the least amount of money?**
    - **Approximately how much money did Paul take altogether on 16th July?**
    - **How many hours is Paul the Baker open for?**
    - **Why do you think Paul is busiest between 10:00 a.m. and 2:00 p.m.?**
- Ask individual children to tell you something else about the bar chart.
- Ask: **By looking at the bar chart, what other information can you tell me about the amount of money Paul the Baker took?**
- Now draw children's attention to the bar line chart on RCM 30.
- Briefly discuss the bar line chart with the children.
- Say: **Look at the bar line chart. What information does it represent?**
- Ask individual children questions similar to the following to assess their ability to extract and interpret information presented in a bar line chart.
    - **Which was the least common item to be dry cleaned?**
    - **14 types of one item were brought in for dry cleaning on 16th July. What type of item was this?**
    - **How many items in total were brought in to be dry cleaned on 16th July?**
    - **How many different categories do Davis Dry Cleaners use to classify their dry cleaning?**
    - **There were 16 'other items' brought in for cleaning. What might some of these 'other items' have been?**
- Ask individual children to tell you something else about the bar line chart.
- Ask: **By looking at the bar line chart, what other information can you tell me about the type of clothing that was brought in for cleaning at Davis Dry Cleaners?**

Success criterion: *Extract and interpret information presented in a line graph*

- Draw children's attention to the line graph on RCM 30.
- Briefly discuss the line graph with the children.
- Say: **Look at the line graph. What information does this line graph show?**
- Ask individual children questions similar to the following to assess their ability to extract and interpret information presented in a line graph.
  - **What is the hottest month in London?**
  - **In which month is the average temperature in London 15 °C?**
  - **For how many months of the year is the average monthly temperature in London below 10 °C?**
  - **What is the difference between the hottest and the coldest months in London?**
  - **Approximately, what is the average temperature for the year in London?**
- Ask individual children to tell you something else about the line graph.
- Ask: **What other information can you find out about the temperature in London using the line graph?**
- Place RCM 30 to one side and show the children RCM 31.
- Draw children's attention to the line graph on RCM 31.
- Briefly discuss the line graph with the children.
- Say: **Look at this line graph. What information is presented in this line graph?**
- Ask individual children questions similar to the following to assess their ability to extract and interpret information presented in a line graph.
  - **How many euros will one pound buy you?**
  - **How many euros will you get for £4?**
  - **How many pounds will you get for €3?**
  - **If something cost €15 how much is this in pounds?**
  - **If something cost £7 how much is this in euros?**
- Ask individual children to tell you something else about the line graph.
- Ask: **What else can you tell me about pounds and euros?**

Success criterion: *Extract and interpret information presented in a pie chart*

- Draw children's attention to the pie chart on RCM 31.
- Briefly discuss the pie chart with the children.
- Say: **Look at this pie chart. What does it show?**
- Ask individual children questions similar to the following to assess their ability to extract and interpret information presented in a pie chart.
  - **Where do most of Travel Bug's customers want to travel to?**
  - **Approximately what per cent of Travel Bug's customers want to travel to Singapore?**
  - **Do more people want to travel to Japan or Australia? Approximately how many per cent more?**
  - **10% of Travel Bug's customers prefer to holiday in two different countries, 5% in one country, 5% in another. Which two countries are these?**
  - **Three countries are equally popular holiday destinations with Travel Bug's customers. Which three countries are these?**
- Ask individual children to tell you something else about the pie chart.
- Ask: **What other things do you know about people's favourite holiday destinations by looking at the pie chart?**

Success criterion: *Extract and interpret information presented in a table*

- Draw children's attention to the table on RCM 31.
- Briefly discuss the table with the children.
- Say: **Look at the table. What information is presented in this table?**
- Ask individual children questions similar to the following to assess their ability to extract and interpret information presented in a table.
  - **How far is it to travel from Milan to Pisa?**
  - **Which two cities are 220 km apart?**
  - **Which two cities have the least amount of distance between them?**
  - **It is further to travel from Rome to Milan than from Rome to Florence. How much further?**
  - **If you were to travel at about 80 km per hour, approximately how long would it take you to travel from Venice to Siena?**
- Ask individual children to tell you something else about the table.
- Say: **Look at the table and tell me anything you notice.**

- Which tables/charts/graphs are easy to interpret information from? Why?
- Which tables/charts/graphs are difficult to interpret information from? Why?
- How did you work out that answer? Which information in the table/chart/graph did you use?
- Ask me a question using the information in the table/chart/graph?

What to do for those children who achieve *above* expectation

- Ask the children more inferential questions about the data, e.g.
  - **July 16th was a Monday. Do you think the bar chart / bar line chart would look different if it were a different day of the week? Why?**
  - **What would happen to this conversion graph if the exchange rate changed between the pound and the euro?**
  - **What would happen to this graph if the euro went up in value against the pound?**
  - **How might the line graph be different for a different city in the UK? A city in Australia?**
  - **Do you think Travel Bug's customers favourite holiday destinations would be the same all year round? Why? Why not?**
  - **How might the road distance table for Italy help you plan a holiday in Italy?**
- Encourage the children to suggest more detailed observations about the data.

What to do for those children who achieve *below* expectation

- Only ask children questions about the following:
  - RCM 30 – Bar chart: Paul the Baker – Amount of money taken on 16th July
  - RCM 30 – Bar line chart: Davis Dry Cleaners – Items brought in for cleaning on 16th July
  - RCM 30 – Line graph: Monthly average temperature in London
  - RCM 31 – Table: Road distances in Italy

Task 27
Handling data

Objective   **NC AT 4**   **NC Levels 4 & 5**
- Describe and interpret results and solutions to problems using the mode, range (Level 4), median and mean (Level 5)

Resources
● large sheet of paper ● marker ● 0–9 die

Task
● Prior to the task, draw the following table on a large sheet of paper.

Die number	Tally	Frequency
0		
1		
2		
3		

> Success criterion: *Understand and work out the mode, range, median and mean*

● Write the following list of numbers on the large sheet of paper: 2, 3, 7, 0, 8, 3 and 5.
● Say: **Look at these numbers, what is the mode?** (3) **How do you know?**
● Remind the children that the term 'mode' refers to the most common or popular value.
● Ask: **What is the range?** (8) **How did you work it out?**
● Remind the children that the 'range' is calculated by finding the difference between the lowest and highest values.
● Ask: **What is the mean?** (4) **How did you work it out?**
● Remind the children that the 'mean' is the equal share value. It is calculated by adding all the values and dividing by the number of values.
● Finally ask: **What is the median?** (3) **How did you work it out?**
● Remind the children that the 'median' is the middle value when all the values are arranged in order. Explain that if there is no middle value, an imaginary value half way between the middle two values is taken.
● Roll a 0–9 die. With the children's help record the results in the table.
● Repeat 25 times.
● Then as a group, work out the mode and calculate the range, median and mean.

● How do you find the mode and range of a set of data?
● How do you calculate the median and mean of a set of data?
● What does the mode tell you about a set of data? What about the range? …median? …mean?

What to do for those children who achieve *above* expectation
● Using the data presented in RCM 30 and RCM 31: Handling data 1 and 2, where appropriate, ask the children to work out the mode, range, median and mean.

What to do for those children who achieve *below* expectation
● Only ask the children to say the mode and range (Level 4).

Self assessment Unit A1

Name _____ Date _____

- I can find the difference between positive and negative numbers

- I can partition, round and order decimals with up to 2 places

- I can use tables facts to work out other facts with decimals

- I can add, subtract, multiply and divide whole numbers and decimals in my head

- I can use a calculator to solve problems involving more than one step

- I can estimate and check the calculations that I do

- I can say whether a number will occur in a sequence, explaining my reason

Self assessment Unit B1

Name _____ Date _____

● I can describe and explain sequences, patterns and relationships ☺ ☺ ☹

● I can suggest hypotheses and test them ☺ ☺ ☹

● I can write and use simple expressions in words and formulae ☺ ☺ ☹

● I can say the squares of numbers to 12 × 12 ☺ ☺ ☹

● I can use tables facts to work out other facts with decimals ☺ ☺ ☹

● I can find pairs of factors ☺ ☺ ☹

● I can recognise a prime number ☺ ☺ ☹

● I can estimate and check the calculations that I do ☺ ☺ ☹

● I can classify 2-D shapes with perpendicular or parallel sides ☺ ☺ ☹

● I can make and draw shapes accurately ☺ ☺ ☹

Collins
New
Primary
Maths

Self assessment Unit C1

Name _____ Date _____

● I can suggest a line of enquiry and plan how to investigate it 　☺ ☺ ☹

● I can answer questions about the data I have represented 　☺ ☺ ☹

● I can represent data in different ways and understand its meaning 　☺ ☺ ☹

● I can work out the mode, range, median and mean 　☺ ☺ ☹

● I can convert from one unit of measure to another 　☺ ☺ ☹

● I can read and interpret scales 　☺ ☺ ☹

Collins New Primary Maths

Self assessment Unit D1

Name _____ Date _____

● I can solve problems with several steps and decide how to carry out the calculation, including using a calculator

● I can add, subtract, multiply and divide whole numbers and decimals in my head

● I can add, subtract and multiply whole numbers and decimals using efficient written methods

● I can estimate the result of a calculation. I know several ways of checking answers

● I can convert from one unit of measure to another

● I know that 1 mile is about 1.6 km, and that 1 km is about $\frac{5}{8}$ of a mile

● I can read and interpret scales

● I can work out the perimeter and area of shapes

Collins New Primary Maths

Self assessment Unit E1

Name _____ Date _____

- I can record the calculations needed to solve a problem and check that my working is correct

- I can work out problems involving fractions, decimals and percentages using a range of methods, including using a calculator

- I can talk about how I solve problems

- I can use place value and my tables to work out multiplication and division facts for decimals

- I can use efficient written methods to add, subtract, multiply and divide whole numbers and decimals

- I can write a large whole number as a fraction of a smaller one

- I can simplify fractions

- I can order a set of fractions

- I can find equivalent fractions

- I can find fractions and percentages of whole numbers

- I can scale up or down to solve problems

Self assessment Unit A2

Name _____ Date _____

● I can explain my reasoning and conclusions, using symbols to represent unknown numbers

● I can solve problems involving more than one step, including using a calculator

● I can use decimals with up to three places and order them on a number line

● I can round decimals to the nearest whole number or the nearest tenth

● I can use tables facts to work out other facts with decimals

● I can add, subtract, multiply and divide whole numbers and decimals in my head

● I can add, subtract, multiply and divide whole numbers and decimals using efficient written methods

● I can estimate and check the result of a calculation

Collins
New
Primary
Maths

Self assessment Unit B2

Name _____ Date _____

● I can describe and explain sequences, patterns and relationships   ☺ ☺ ☹

● I can suggest hypotheses and test them   ☺ ☺ ☹

● I can write and use simple expressions in words and formulae   ☺ ☺ ☹

● I can use a table to help me solve a problem   ☺ ☺ ☹

● I can identify and record what I need to do to solve a problem, checking my answer makes sense and is accurate   ☺ ☺ ☹

● I can say the squares of numbers to 12 × 12 and work out the squares of multiples of 10   ☺ ☺ ☹

● I can use tables facts to work out related facts with decimals   ☺ ☺ ☹

● I can work out which numbers less than 100 are prime   ☺ ☺ ☹

● I can estimate and check the result of a calculation   ☺ ☺ ☹

● I can use the properties of parallel and perpendicular sides to describe and classify 2-D shapes   ☺ ☺ ☹

● I can make and draw shapes accurately   ☺ ☺ ☹

● I can use co-ordinates   ☺ ☺ ☹

● I can reflect shapes on grids   ☺ ☺ ☹

Collins
New
Primary
Maths

Self assessment Unit C2

Name _____ Date _____

- I can use data to solve problems ☺ ☺ ☹

- I can represent data in different ways and understand its meaning ☺ ☺ ☹

- I can work out the mode, range, median and mean ☺ ☺ ☹

- I can use the language of chance and likelihood ☺ ☺ ☹

- I can convert measures between units including decimals ☺ ☺ ☹

- I know that 1 lb is just over 450 g, and that 1 kg is about $2\frac{1}{4}$ lb ☺ ☺ ☹

- I can read and interpret scales ☺ ☺ ☹

- I can compare readings from different scales ☺ ☺ ☹

- I can use a calculator to solve problems involving more than one step ☺ ☺ ☹

- I can represent data in different ways and understand its meaning ☺ ☺ ☹

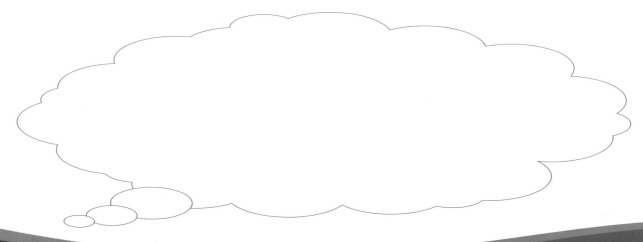

Self assessment Unit D2

Name _____ Date _____

- I can solve problems with several steps and decide how to carry out the calculation ☺ 😐 ☹

- I can use a calculator to solve problems with several steps ☺ 😐 ☹

- I can add, subtract, multiply and divide whole numbers and decimals in my head ☺ 😐 ☹

- I can add, subtract, multiply and divide whole numbers and decimals using efficient written methods ☺ 😐 ☹

- I can estimate and check the result of a calculation ☺ 😐 ☹

- I can convert one measurement to another using a related unit. I use decimals to do this ☺ 😐 ☹

- I can estimate angles, and use a protractor to measure and draw them ☺ 😐 ☹

- I know the angle sum of a triangle is 180° and the sum of angles around a point is 360° ☺ 😐 ☹

- I can use co-ordinates ☺ 😐 ☹

- I can rotate shapes on grids ☺ 😐 ☹

- I can translate shapes on grids ☺ 😐 ☹

Collins
New
Primary
Maths

Self assessment Unit E2

Name _____ Date _____

● I can record the calculations needed to solve a problem and check that my working is correct   ☺ 😐 ☹

● I can use a table to help me solve a problem   ☺ 😐 ☹

● I can talk about how I solve problems   ☺ 😐 ☹

● I can describe and explain sequences, patterns and relationships   ☺ 😐 ☹

● I can suggest hypotheses and test them   ☺ 😐 ☹

● I can write and use simple expressions in words and formulae   ☺ 😐 ☹

● I can work out problems involving fractions, decimals and percentages using a range of methods   ☺ 😐 ☹

● I can write a larger whole number as a fraction of a smaller one   ☺ 😐 ☹

● I can simplify fractions   ☺ 😐 ☹

● I can order a set of fractions   ☺ 😐 ☹

● I can find fractions and percentages of whole numbers   ☺ 😐 ☹

● I can work out a quantity as a percentage of another   ☺ 😐 ☹

● I can find equivalent percentages, decimals and fractions   ☺ 😐 ☹

● I can solve problems using ratio and proportion   ☺ 😐 ☹

Collins
New
Primary
Maths

Self assessment Unit A3

Name _____ Date _____

- I can solve problems involving more than one step.

- I can explain the reason for my choice of method and say whether I think it was effective

- I can use a calculator to solve problems involving more than one step

- I can use decimals with up to three places

- I can partition, round and order decimals with up to three places

- I can add, subtract, multiply and divide whole numbers and decimals in my head

- I can use efficient written methods to add, subtract, multiply and divide integers and decimal numbers

- I can estimate and check the result of a calculation

Collins
New
Primary
Maths

Self assessment Unit B3

Name _____ Date _____

● I can use a table to help me solve a problem ☺ ☺ ☹

● I can identify and record what I need to do to solve a problem, checking that my answer makes sense and is accurate ☺ ☺ ☹

● I can use a calculator to solve problems involving more than one step ☺ ☺ ☹

● I can describe and explain sequences, patterns and relationships ☺ ☺ ☹

● I can suggest hypotheses and test them ☺ ☺ ☹

● I can write and use simple expressions in words and formulae ☺ ☺ ☹

● I can say the squares of numbers to 12 × 12 and work out the squares of multiples of 10 ☺ ☺ ☹

● I can use tables facts to work out related facts with decimals ☺ ☺ ☹

● I can work out which numbers less than 100 are prime ☺ ☺ ☹

● I can estimate and check the result of a calculation ☺ ☺ ☹

● I can identify 3-D solids with perpendicular or parallel edges or faces ☺ ☺ ☹

● I can make and draw shapes accurately ☺ ☺ ☹

Collins New Primary Maths

Self assessment Unit C3

Name _____ Date _____

● I can collect and present data in a variety of ways and use my results to solve problems

● I can represent data in a variety of ways and answer questions about the data including interpreting pie charts

● I can use the different averages to solve problems

● I can use the language of chance and likelihood

● I can convert measures between units including decimals

● I can read and interpret scales

● I can compare readings from different scales   ☺ ☺ ☹

● I can solve problems involving more than one step   ☺ ☺ ☹

Self assessment Unit D3

Name _____ Date _____

● I can solve problems with several steps and decide how to carry out the calculation, including using a calculator   ☺ ☺ ☹

● I can add, subtract, multiply and divide whole numbers and decimals in my head   ☺ ☺ ☹

● I can add, subtract, multiply and divide whole numbers and decimals using efficient written methods   ☺ ☺ ☹

● I can estimate the result of a calculation. I know several ways of checking answers   ☺ ☺ ☹

● I can convert one measurement to another using a related unit. I use decimals to do this   ☺ ☺ ☹

● I know that 1 pint is just over half a litre, and that 1 litre is about $1\frac{3}{4}$ pints   ☺ ☺ ☹

● I can read and interpret scales   ☺ ☺ ☹

● I can compare readings from different scales   ☺ ☺ ☹

● I can find the perimeter and area of shapes   ☺ ☺ ☹

Collins New Primary Maths

Self assessment Unit E3

Name _____ Date _____

● I can record the calculations needed to solve a problem and check that my working is correct ☺ ☺ ☹

● I can work out problems involving fractions, decimals and percentages using a range of methods ☺ ☺ ☹

● I can use place value and my tables to work out multiplication and division facts ☺ ☺ ☹

● I can use standard written methods to add, subtract, multiply and divide whole numbers and decimals ☺ ☺ ☹

● I can write a large whole number as a fraction of a smaller one ☺ ☺ ☹

● I can simplify fractions ☺ ☺ ☹

● I can order a set of fractions ☺ ☺ ☹

● I can find fractions and percentages of whole numbers ☺ ☺ ☹

● I can work out a quantity as a percentage of another ☺ ☺ ☹

● I can find equivalent percentages, decimals and fractions ☺ ☺ ☹

● I can solve problems using ratio and proportion ☺ ☺ ☹

Test 1
Mental mathematics test questions and answers

Say: **For this group of questions, you will have 5 seconds to work out each answer and write it down.**

	The questions	Answers
1.	Add fifty to eight hundred and seventy.	920
2.	Multiply seven by nine.	63
3.	What is one hundred less than seven thousand?	6900
4.	How many sides has a pentagon?	5
5.	Write two point six three to the nearest whole number.	3

Say: **For the next group of questions, you will have 10 seconds to work out each answer and write it down.**

		Answers
6.	What temperature is ten degrees warmer than minus two degrees Celsius?	8 °C
7.	What is the total of seventy, fifty and thirty?	150
8.	What is four hundred and six add five hundred and three?	909
9.	Add seven to four point three.	11·3
10.	Write down the fractional equivalent of nought point seven.	$\frac{7}{10}$
11.	Imagine a cuboid. How many vertices does it have?	8
12.	How many millimetres are there in twelve centimetres?	120 mm
13.	Subtract one point eight from four.	2·2
14.	Add one and a half to four and a quarter.	$5\frac{3}{4}$
15.	Each side of a hexagon is six centimetres. What is its perimeter?	36 cm

Say: **For the next group of questions, you will have 15 seconds to work out each answer and write it down.**

		Answers
16.	One note pad costs one pound and sixty-five pence. How much do seven note pads cost?	£11.55
17.	Multiply thirty by sixty.	1800
18.	Add together six and nine, then multiply the result by three.	45
19.	What number is thirty-seven less than seventy-six?	39
20.	Tickets for the circus are: adults twelve pounds and children nine pounds. How much would it cost for two adults and three children?	£51

Say: **Now put down your pencil. The test is finished.**

Test 1
Papers A and B answers

Paper A

1. 124, 142, 214, 241, 412, 421

2. 125, 146

3.

70	71	(72)	73	74
75	76	77	78	79
(80)	81	82	83	84
85	86	87	(88)	89
90	91	92	93	94
95	(96)	97	98	99

4.

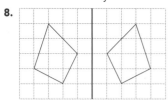

5. 619

6. 6

7. Check the accuracy of children's drawing

8.

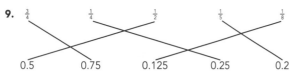

9.

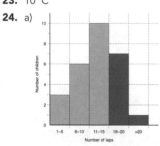

10. 168 km

11. a) 36   b) 28

12. 64 cm²

13. 4 kg 800 g

14. 40%

15. 22 min

16. 5

17. 17 696

18. 6·77

19. a) £34.85   b) £44.35

20. 44 cm

21. a) $\frac{1}{3}$ b) Accept £125 to £145

22. a) (2, 5)   b) (8, 9)

23. 10 °C

24. a)

(bar chart: Number of children vs Number of laps)

a) $\frac{1}{3}$

25. a) Thursday   b) 4

26. a) 48 km   b) 14 km

Paper B

1. 5, 8, 11

2. a) 384   b) 414   c) 8

3. 4·573, 4·7, 5·354, 5·374, 5·43

4.

	Answer to the nearest whole number
35 712 ÷ 45	794
38·76 × 27	1047
789·7 − 463·82	326
79·83 − (93·1 − 34·39)	21

5. a) 8   b) 16

6. 3, 9, 81

7. −9, −4, −1, 3, 6, 8

8. 511

9. a) 10 hr   b) 67 hr

10. 55 km per hour

11.

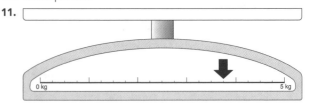

12. 167

13. 15:35

14. a) £726.45   b) £8.60

15. a) 7   b) 9·06

16. A = 24°   B = 73°

17. Jumo: 9   Tom: 12

18. −20, −55

19. a) £22.72   b) 6p

20. 136·3 kg

21. 108

22.

	Has no right angle	Has four equal angles	Has one acute angle
Has one pair of perpendicular sides			D
Has two pairs of perpendicular sides	B	A C	

23.

24. a) 0·7 litres   b) 1·3 litres

25.

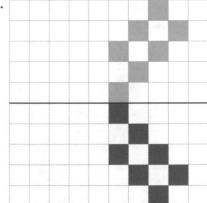

Mental mathematics test

Name ....

Date .... Class ....

Total marks [ ]

Time: 5 seconds

| 1 | | 50   870 | ___ 1 |

| 2 | | | ___ 2 |

| 3 | | 7000 | ___ 3 |

| 4 | | | ___ 4 |

| 5 | | 2·63 | ___ 5 |

| 11 | | | ___ 11 |

| 12 | | mm | 12 cm | ___ 12 |

| 13 | | 4   1·8 | ___ 13 |

| 14 | | $1\frac{1}{2}$ $4\frac{1}{4}$ | ___ 14 |

| 15 | | cm | 6 cm | ___ 15 |

Time: 10 seconds

| 6 | | °C | –2 °C | ___ 6 |

| 7 | | 70 50 30 | ___ 7 |

| 8 | | 406   503 | ___ 8 |

| 9 | | 7   4·3 | ___ 9 |

| 10 | | 0·7 | ___ 10 |

Time: 15 seconds

| 16 | £ | £1.65 | ___ 16 |

| 17 | | 30   60 | ___ 17 |

| 18 | | 6   9   3 | ___ 18 |

| 19 | | 37   76 | ___ 19 |

| 20 | £ | £12   £9 | ___ 20 |

Year 6 Test 1

MATHEMATICS

YEAR 6

TEST 1
PAPER A

LEVELS 3–5

CALCULATOR
NOT ALLOWED

PAGE	MARKS
3	
4	
5	
6	
7	
8	
9	
10	
11	
12	
13	
TOTAL	

RESOURCES

• pencil
• ruler
• protractor

Name

Date

Class

Instructions

You may not use a calculator to answer any questions in this test.

Work as quickly and as carefully as you can.

You have 45 minutes for this test.

If you cannot do one of the questions, go on to the next one.

You can come back to it later if you have time.

If you have finished before the end, go back and check your work.

Follow the instructions for each question carefully.

 This shows where you need to put your answer.

If you need to do any working out, you can use any space on the page.

Some questions have an answer box like this:

Show your working. You may get a mark.

For these questions you may get a mark for showing your working.

1 Arrange the following numbers in order, smallest to largest.

241   421   142   214   124   412

smallest

2 Complete the following sequence.

41   62   83   104

3 Circle the numbers divisible by 8.

70	71	72	73	74
75	76	77	78	79
80	81	82	83	84
85	86	87	88	89
90	91	92	93	94
95	96	97	98	99

4 Write these numbers in the correct places in the diagram.

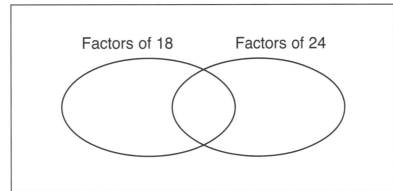

4   6   9   10

Factors of 18   Factors of 24

4

2 marks

5 Calculate 1008 − 389

5

1 mark

6 Write the missing number in the box.

          1 7 4 5

×   ▢

——————————————

1 0 4 7 0

6

1 mark

7

Here is a sketch of a triangle.

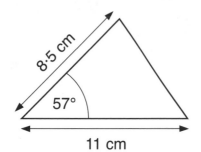

Triangle is not drawn to scale.

Draw this triangle accurately using a protractor and a ruler.

_____
2 marks

8

Reflect this shape in the mirror line.

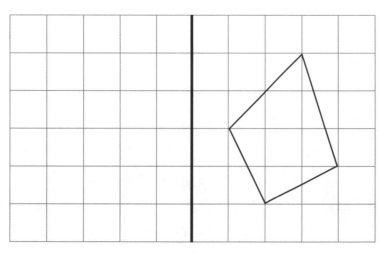

_____
2 marks

9 Draw a line between each fraction and its equivalent decimal.
One has been done for you.

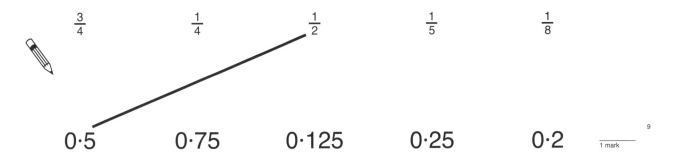

$\frac{3}{4}$ $\frac{1}{4}$ $\frac{1}{2}$ $\frac{1}{5}$ $\frac{1}{8}$

0·5   0·75   0·125   0·25   0·2

9

1 mark

10 Ahmed lives 1·4 km from his school. He walks there and back each day. If he goes to school 5 days a week, how far does he walk in a 12-week term?

Show your working. You may get a mark.

10

2 marks

11 a) Three times as many girls play hockey as boys. If 27 girls are playing hockey, how many children in total are playing hockey?

11a

1 mark

b) Twice as many boys play football as girls. If 42 children are playing football, how many boys are playing football?

11b

1 mark

12 Gina has some rectangular tiles like this.

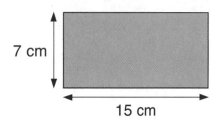

7 cm

15 cm

This diagram is not to scale.

She makes a pattern from them.

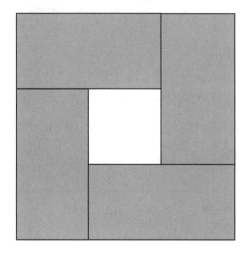

What is the area of the hole in the middle of the pattern?

12

2 marks

13 A box of baked beans holds 12 tins. Each tin contains 400 g of baked beans. What is the total weight of baked beans in the box?

13

1 mark

14 What percentage of this square is shaded?

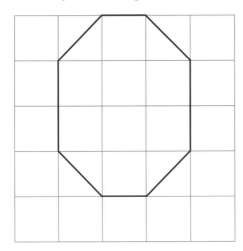

$\dfrac{}{\text{1 mark}}$ 14

15 David's train leaves at 6 o'clock. How long to go before it leaves?

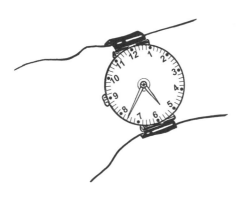

$\dfrac{}{\text{1 mark}}$ 15

16 Rosie wants to buy 1·4 litres of lemonade. The lemonade comes in tins containing 330 ml. How many tins must she buy?

$\dfrac{}{\text{1 mark}}$ 16

17 Calculate 632 × 28

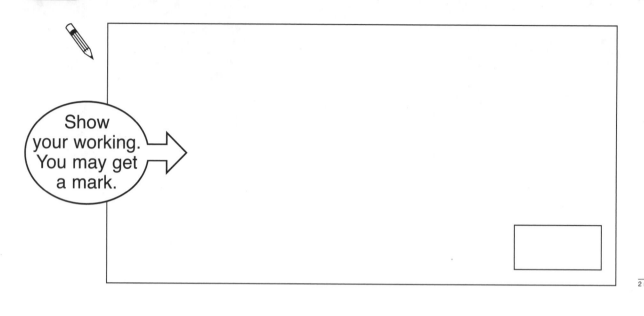

Show your working. You may get a mark.

17

2 marks

18 These are the points that five friends scored playing a game. Circle the score that is nearest to 7.

Sara	Pervais	Nick	Joseph	Ana
6·57	6·77	7·73	7·51	7·28

18

1 mark

19 In Joe's Music Store, CDs normally cost £9.90. During the sale, the price of one CD is reduced to £4.95, or you can buy five CDs for £20.

Parin buys 8 CDs in the sale.

a) How much money does he pay for them?

19a

1 mark

b) How much money does he save buying them in the sale?

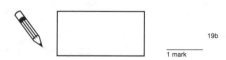

19b

1 mark

20 Calculate the perimeter of this shape.

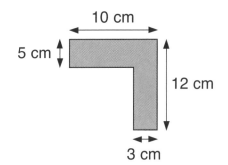

5 cm

10 cm

12 cm

3 cm

This diagram is not to scale.

Show your working. You may get a mark.

cm

20

2 marks

21 A school fair raised £850. The pie chart shows how the money was raised.

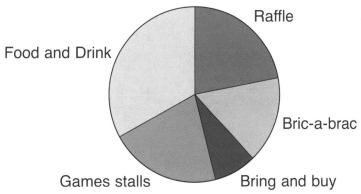

Raffle

Food and Drink

Bric-a-brac

Games stalls

Bring and buy

a) Estimate the fraction of the total that was raised by Food and Drink.

21a

1 mark

b) Estimate how much was raised by Bric-a-brac.

21b

1 mark

22 What are the co-ordinates of points A and B?

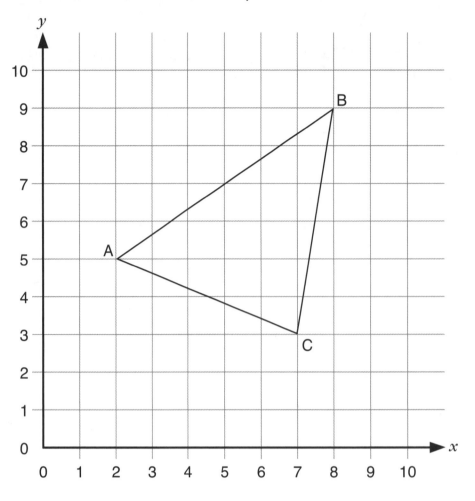

Point A ✏️ [   ]
22a
1 mark

Point B ✏️ [   ]
22b
1 mark

23 Calculate the temperature difference between London and Rome.

London

Rome

✏️ [   ]
23
1 mark

24 The table below shows the number of laps that children swam in a sponsored swim.

Number of laps	1–5	6–10	11–15	16–20	> 20
Number of children	3	6	10	7	1

a) Complete the chart below.

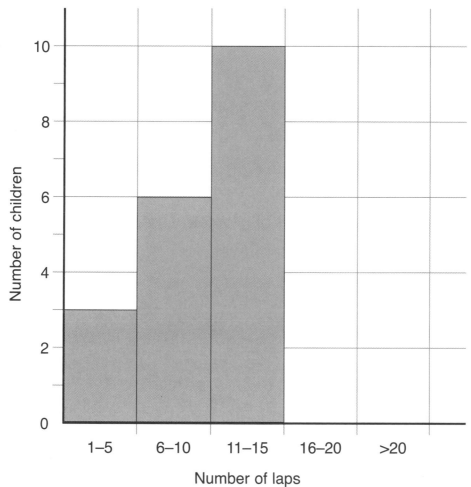

Number of children

Number of laps

b) What fraction of children swam 10 or fewer lengths?

25 This table shows the birds that Sue saw in her garden during one week.

	Mon	Tue	Wed	Thur	Fri	Sat	Sun
Sparrow	✓			✓		✓	
Robin		✓		✓			✓
Starling	✓	✓	✓		✓	✓	✓
Blackbird			✓	✓		✓	
Pigeon		✓			✓		

a) On which day did Sue see both a robin and a blackbird?

25a

1 mark

b) On how many days did Sue see less than three different types of bird?

25b

1 mark

26 This graph shows the number of kilometres per litre of petrol that a large car and a small car travel.

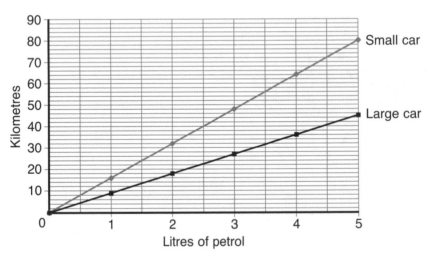

a) How far will a small car travel on 3 litres of petrol?

26a

1 mark

b) How much further will a small car travel on 2 litres of petrol than a large car?

26b

1 mark

End of test

Total out of 4 _____

MATHEMATICS

YEAR 6

TEST 1 PAPER B **LEVELS 3–5**

CALCULATOR ALLOWED

PAGE	MARKS
3	
4	
5	
6	
7	
8	
9	
10	
11	
12	
TOTAL	

RESOURCES
- pencil
- ruler
- calculator

Name

Date

Class

Instructions

You may use a calculator to answer any questions in this test.

Work as quickly and as carefully as you can.

You have 45 minutes for this test.

If you cannot do one of the questions, go on to the next one.

You can come back to it later if you have time.

If you have finished before the end, go back and check your work.

Follow the instructions for each question carefully.

 This shows where you need to put your answer.

If you need to do any working out, you can use any space on the page.

Some questions have an answer box like this:

For these questions you may get a mark for showing your working.

1 Tim makes a sequence of numbers by adding the same number each time. Fill in the missing numbers in the sequence.

2				14

1 mark

1

2 Write in the missing numbers.

a) $637 - \boxed{} = 253$

1 mark

2a

b) $\boxed{} \div 9 = 46$

1 mark

2b

c) $\boxed{} \times 54 = 432$

1 mark

2c

3 Order these numbers from smallest to largest.

5·354   4·7   5·374   4·573   5·43

smallest

1 mark

3

4 Write the answer to these calculations, rounded to the nearest whole number.

	Answer to the nearest whole number
35 712 ÷ 45	
38·76 × 27	
789·7 − 463·82	
79·83 − (93·1 − 34·39)	

2 marks

4

5 a) Peter thinks of a number. He doubles the number and then subtracts 27. The answer is −11. What number did Peter think of?

1 mark

5a

b) Sam thinks of a number. He multiples it by 5 then divides the product by 16. The answer is 5. What was Sam's number?

1 mark

5b

6 Each number in this sequence is the square of the number before it. Complete the sequence.

6561

1 mark

6

7 Arrange these numbers in order, smallest to largest.

$$-9 \qquad 8 \qquad 3 \qquad -4 \qquad -1 \qquad 6$$

[ ] [ ] [ ] [ ] [ ] [ ]

7

1 mark

smallest

8 Calculate 35% of 1460.

[ ]

8

1 mark

9 These are the opening hours of a bakery.

	Opening hours
Monday	7 a.m. to 6 p.m.
Tuesday	7 a.m. to 6 p.m.
Wednesday	7 a.m. to 1 p.m.
Thursday	8 a.m. to 6 p.m.
Friday	7 a.m. to 7 p.m.
Saturday	6 a.m. to 5 p.m.
Sunday	6 a.m. to 12 p.m.

a) How many hours is the bakery open
   on Thursday?

[ ]

9a

1 mark

b) For how many hours a week is the
   bakery open?

[ ]

9b

1 mark

10 Rita drives for 4 hours in the morning and covers 230 km. She then stops for lunch. After lunch she drives another 3 hours and covers 155 km.

What was the average speed for her journey that day?

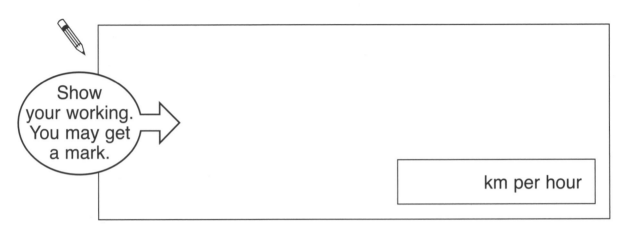

Show your working. You may get a mark.

| km per hour |

10

2 marks

11 Draw an arrow on the scale so it shows a weight of 3·75 kg.

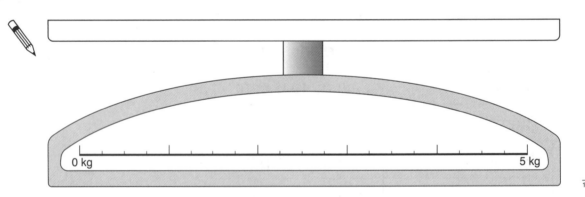

0 kg   5 kg

11

1 mark

12 What is the value of $12y - 37$ if $y = 17$?

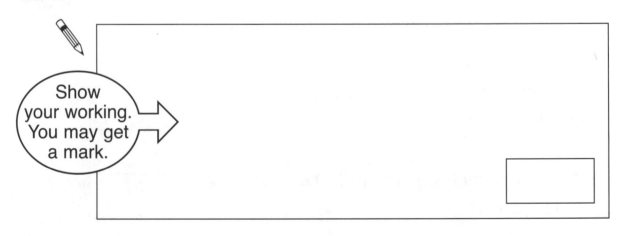

Show your working. You may get a mark.

12

2 marks

13

I'm going swimming at twenty-five to four this afternoon.

Show this time on the
24-hour digital clock below.

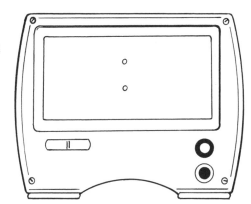

13

1 mark

14   a) 167 children go on an outing to the zoo. Entrance tickets
cost £4.35 per child. How much do the tickets for all the
children cost?

14a

1 mark

b) 27 adults accompany the children. The tickets for these
adults cost £232.20 altogether. How much does each adult
ticket cost?

14b

1 mark

15   Write in the missing numbers.

a) $67\cdot8 \times$ ☐ $= 474\cdot6$

15a

1 mark

b) $56\cdot2 - (134\cdot64 - 87\cdot5) =$

15b

1 mark

16 How many degrees are angles A and B?

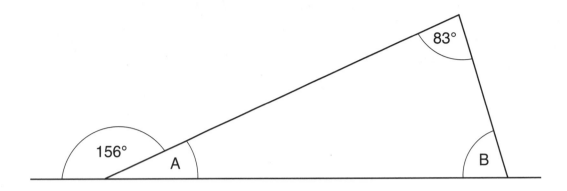

A    B

16

2 marks

17 Jumo and Tom have 21 marbles between them. Jumo has 3 fewer marbles than Tom. How many marbles do Jumo and Tom each have?

Jumo   Tom

17

1 mark

18 Each number in the sequence is 35 less than the number before it. Write in the next two numbers.

| 85 | 50 | 15 | | |

18

1 mark

Total out of 4 _____

19 Jemma went to the DIY store. She bought a hammer costing £8.63, a saw for £12.89 and a box of screws for £5.76.

a) How much change does she get from a £50 note?

19a

1 mark

b) There are 96 screws in the box of screws that Jemma buys. How much does each screw cost?

19b

1 mark

20 There are 54 large bags of sugar and 79 small bags of sugar on the shelf in a shop. The large bags contain 1·5 kg and the small bags contain 700 g.

What is the total weight of sugar on the shelf?

Show your working. You may get a mark.

kg

20

2 marks

21 Calculate $\frac{3}{7}$ of 252.

21

1 mark

22 Look at the four quadrilaterals below.

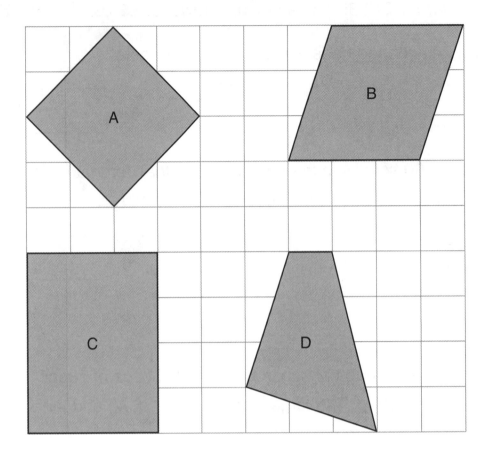

Write the letter of each quadrilateral in one area of the table below.

	Has no right angle	Has four equal angles	Has one acute angle
Has one pair of perpendicular sides			
Has two pairs of parallel sides			

22

2 marks

23 This is a pack of 20 cards numbered from 1 to 20.

Draw an **X** on the scale below to show the probability of choosing a number greater than 15.

0           $\frac{1}{2}$ 1

23

1 mark

24 Bindi has measured out some milk for a recipe.

a) How much milk is she using?

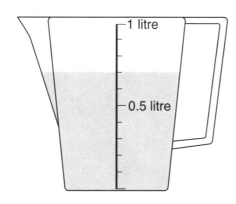

litres

24a

1 mark

b) If Bindi poured the milk from a full 2-litre carton, how much milk is still left in the carton?

litres

24b

1 mark

25 Reflect this pattern about the mirror line.

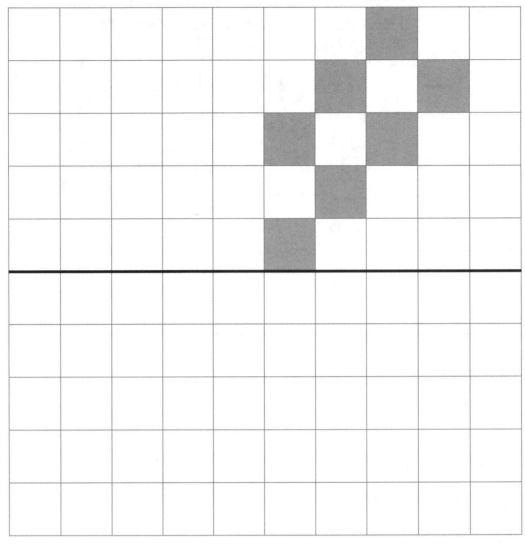

25

2 marks

End of test

Total out of 2 _____

Test 2
Mental mathematics test questions and answers

Say: **For this group of questions, you will have 5 seconds to work out each answer and write it down.**

	The questions	Answers
1.	What is half of seventy?	35
2.	Write in figures the number nine thousand and twenty.	9020
3.	How many minutes are there in three-quarters of an hour?	45 min
4.	Subtract seventy from three thousand.	2930
5.	How many grams are there in eight kilograms?	8000 g

Say: **For the next group of questions, you will have 10 seconds to work out each answer and write it down.**

		Answers
6.	What fraction of five pounds is fifty pence?	$\frac{1}{10}$
7.	A rectangle measures six centimetres by eleven centimetres. What is its area?	66 cm^2
8.	What is the sum of seventy and eight hundred and fifty?	920
9.	Write down the decimal equivalent of seventy-one per cent.	0·71
10.	What is three times two hundred and thirty?	690
11.	Multiply four point two by three.	12·6
12.	What is three-quarters of one hundred and sixty?	120
13.	What is thirty multiplied by thirty?	900
14.	Imagine a triangular prism. How many edges does it have?	9
15.	What is ten per cent of three hundred and ten?	31

Say: **For the next group of questions, you will have 15 seconds to work out each answer and write it down.**

		Answers
16.	What is six point four subtract two point seven.	3·7
17.	Multiply thirty-six by eight.	288
18.	Two angles in a triangle are thirty-five degrees and fifty degrees. What is the third angle?	95°
19.	Divide five point four by six.	0·9
20.	A sale has forty per cent off. Carl buys an item that cost twenty-five pounds before the sale. How much money did he save?	£10

Say: **Now put down your pencil. The test is finished.**

Test 2
Papers A and B answers

Paper A

1. 445 and 470

2. 179, 719, 791, 917, 971, 1079

3. 7

4. 121

5. y = 8 or –22 because the difference between 8 and –7 is 15 and the difference between –7 and –22 is also 15.

6. 94

7. $47\frac{3}{17}$ or 47.18

8. a) Check that numbers written in boxes B, C and D meet the given criteria.

   b) Box A is always empty because a square number can never be a prime number because all square numbers are composite numbers (i.e. they have more than two factors).

9. –4 °C

10. John: 12 km   Gita: 3 km

11. a) Sept / Oct   b) April / Oct

12. 4·8 km

13. 367

14. £1.46

15. 1·007 m²

16.
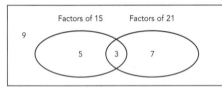

17. 35%

18. a) 30   b) 45

19. 60

20. 72 cm

21. 28, 42, 63

22. a) 1·3, 1·9 km   b) 1·5 km

23. 9:34 p.m.

24. A = 38°   B = 74°

25.
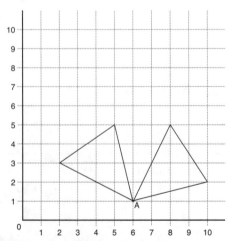

Paper B

1. 5·6

2. $\frac{9}{12}$ or $\frac{12}{16}$

3. a) $2\frac{1}{2}$ b) £14.25

4.
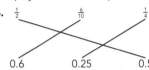

5. 768, 3072

6. 36

7. 8

8. a) £1192.50   b) 94

9. 7 × 11 × 29

10. 35 min

11. 800 g

12. 4, 5, 6, 7

13. 426, 427, 462, 467, 472, 476

14. 5.476, 5.564, 5.654, 5.674, 5.764

15. a) 65   b) 20

16.
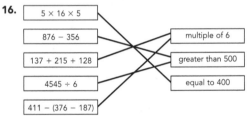

17. 12 cm²

18.
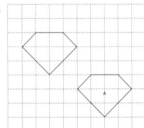

19. 100 g

20. 147°

21.
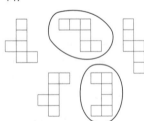

22. (5, 0), (7, 2), (7,4)

23.

24. a) 44   b) Bus

25. a) 1·38 kg   b) 33p

Mental mathematics test

Name ....

Date .... Class ....

Total marks [ ]

Time: 5 seconds

| 1 | | 70 | _____ 1 |

| 2 | | | _____ 2 |

| 3 | min | $\frac{3}{4}$ hr | _____ 3 |

| 4 | | 3000 70 | _____ 4 |

| 5 | g | 8 kg | _____ 5 |

| 11 | | 4·2 | _____ 11 |

| 12 | | 160 | _____ 12 |

| 13 | | 30 | _____ 13 |

| 14 | | | _____ 14 |

| 15 | | 310 | _____ 15 |

Time: 10 seconds

| 6 | | £5 50p | _____ 6 |

| 7 | cm² | | _____ 7 |

| 8 | | 70 850 | _____ 8 |

| 9 | | 71% | _____ 9 |

| 10 | | 230 | _____ 10 |

Time: 15 seconds

| 16 | | 6·4 2·7 | _____ 16 |

| 17 | | 36 8 | _____ 17 |

| 18 | ° | 35° 50° | _____ 18 |

| 19 | | 5·4 6 | _____ 19 |

| 20 | £ | 40% £25 | _____ 20 |

Year 6 Test 2

MATHEMATICS

YEAR 6

TEST 2
PAPER A

LEVELS 3–5

CALCULATOR
NOT ALLOWED

PAGE	MARKS
3	
4	
5	
6	
7	
8	
9	
10	
11	
12	
13	
TOTAL	

RESOURCES
- pencil
- ruler
- protractor

Name

Date

Class

Instructions

You may not use a calculator to answer any questions in this test.

Work as quickly and as carefully as you can.

You have 45 minutes for this test.

If you cannot do one of the questions, go on to the next one.

You can come back to it later if you have time.

If you have finished before the end, go back and check your work.

Follow the instructions for each question carefully.

 This shows where you need to put your answer.

If you need to do any working out, you can use any space on the page.

Some questions have an answer box like this:

Show your working. You may get a mark.

For these questions you may get a mark for showing your working.

1 Here is a number line. Write the missing numbers in the boxes.

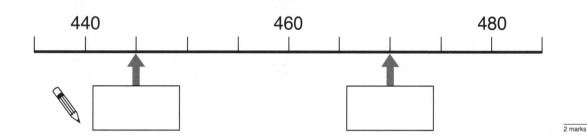

2 Arrange the following numbers in order, smallest to largest.

917   791   179   971   1079   719

smallest

3 Write the missing number in the box.

$2975 \times \boxed{} = 20\,825$

4 Write the next square number after 100.

4

1 mark

5 The number in the box is the difference between the two numbers in the circles joined to it.

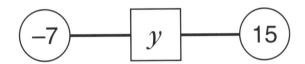

y can have two correct values

Is Tom right?

Explain your answer.

5

2 marks

6 Write the next number in the following sequence.

34   49   64   79

6

1 mark

7 Calculate 802 ÷ 17

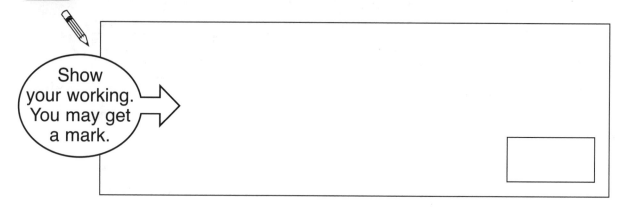

Show your working. You may get a mark.

2 marks

7

8 a) Write a number less than 100 in each of Boxes B, C and D.

	Square number	Not a square number
Prime number	A	B
Not a prime number	C	D

8a

1 mark

b) Explain why Box A will always be empty.

8b

1 mark

9 The temperature in London is 5 °C. It is 9° colder in Oslo.

What is the temperature in Oslo?

1 mark

9

10

John cycled 11·72 km in an hour. His friend Gita walked 3·45 km in the same time. How far did each of them travel, rounded to the nearest whole kilometre?

John ✎ [   ]   Gita ✎ [   ]

10
1 mark

11

The line graph below shows the maximum temperatures during the year in central Italy.

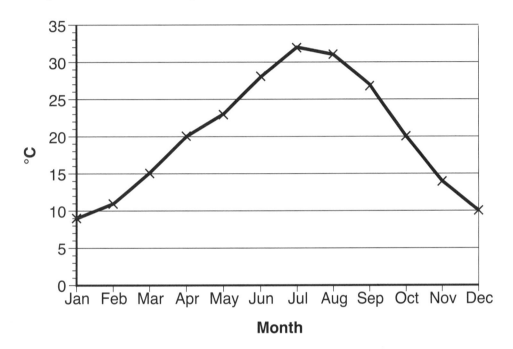

Month

a) During which 2-month period does the maximum temperature change the most?

✎ [   ]

11a
1 mark

b) In which 2 months of the year is the maximum temperature the same?

✎ [   ]

11b
1 mark

12 Amy runs 12 laps of a 400 m racetrack. How far does she run?

km

1 mark

12

13 Calculate 743 − 376

1 mark

13

14 Kim bought 6 tins of tomatoes costing 38p each, and a dozen eggs costing £1.26. How much change did she get from £5?

Show your working. You may get a mark.

2 marks

14

15 Calculate the area of this shape.

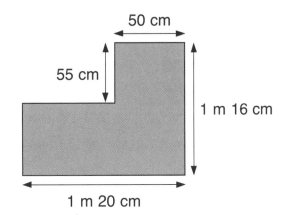

50 cm

55 cm

1 m 16 cm

1 m 20 cm

This diagram is
not to scale.

Show
your working.
You may get
a mark.

m²

2 marks

16 Write these numbers in the correct places in the diagram.

3   5   7   9

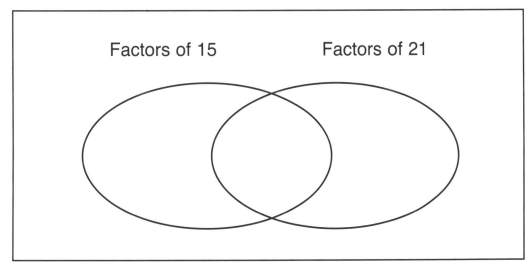

Factors of 15   Factors of 21

2 marks

17   Anna knocked over a 1·5 kg bag of flour and spilt some.
The flour that was left in the bag weighed 965 g. Circle the
percentage that is nearest to the percentage of flour that
she spilt.

   25%   35%   45%   55%   65%

18   The pie chart below shows information about children's
favourite fruit.

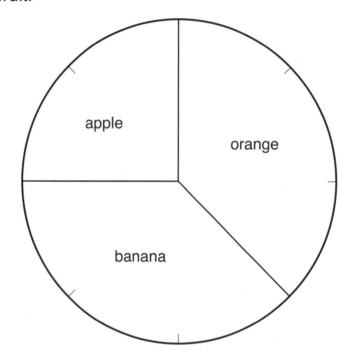

Altogether 120 children were asked about their favourite fruits.

a) How many of the 120 children said they preferred apples?

b) How many of the 120 children said they preferred bananas?

19 A group of children set off on a run. Only 4 out of every 5 children completed it. How many children set off on the run, if 12 children did not complete it?

Show your working. You may get a mark.

19

2 marks

20 The shape below is made up of 5 squares.

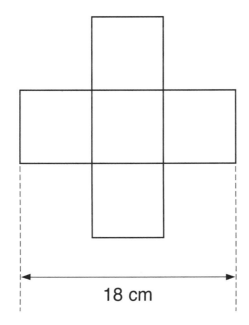

18 cm

What is its perimeter?

20

2 marks

21 Circle all the multiples of 7.

18   28   36   42   55   63

1 mark

22 Louis walks every day for two weeks.

This is how far he walks each day.

1·3 km	1·8 km	1·6 km	1·4 km	1·9 km	1·5 km	1·4 km
1·9 km	1 km	1·1 km	1·3 km	1·3 km	1·9 km	1·6 km

a) What is the mode distance?

22a
1 mark

b) What is his mean (average) distance?

22b
1 mark

Total out of 3 _____

23 Tom catches a train at 5:27 p.m. The journey lasts 3 hours 45 minutes and it takes him another 22 minutes to walk from the station to his house. At what time does Tom arrive home?

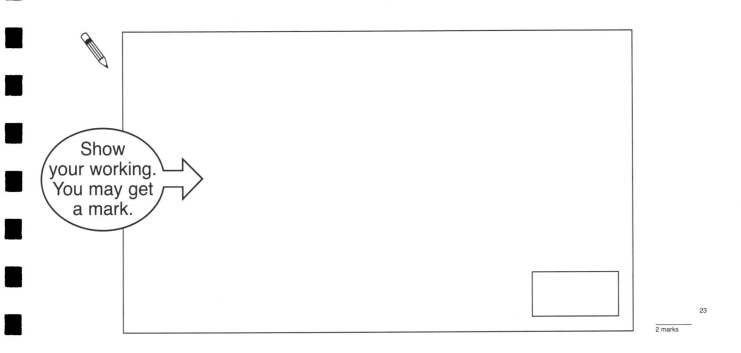

Show your working. You may get a mark.

23

2 marks

24 Measure angles A and B.
Use an angle measurer (protractor).

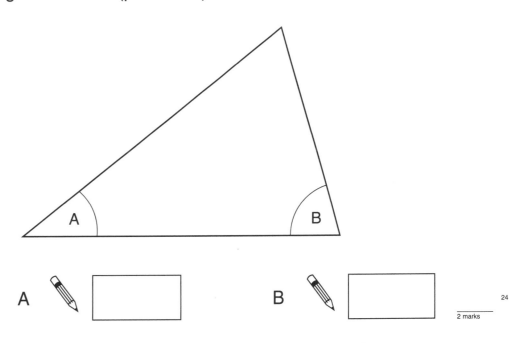

A [ ]

B [ ]

24

2 marks

Rotate this triangle 90° clockwise about point A.

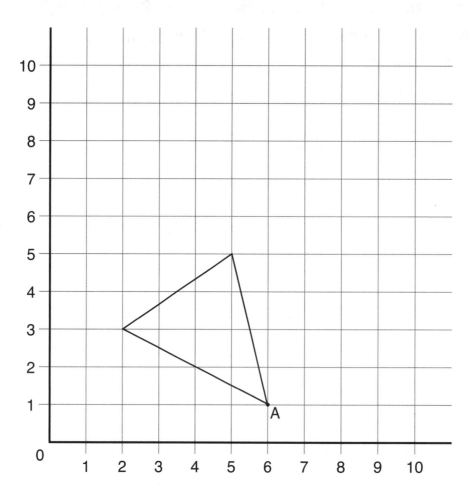

25

2 marks

End of test

Total out of 2 _____

MATHEMATICS

YEAR 6

TEST 2 PAPER B | **LEVELS 3–5**

CALCULATOR ALLOWED

PAGE	MARKS
3	
4	
5	
6	
7	
8	
9	
10	
11	
12	
13	
TOTAL	

RESOURCES

- pencil
- ruler
- calculator

Name

Date

Class

Instructions

You may use a calculator to answer any questions in this test.

Work as quickly and as carefully as you can.

You have 45 minutes for this test.

If you cannot do one of the questions, go on to the next one.

You can come back to it later if you have time.

If you have finished before the end, go back and check your work.

Follow the instructions for each question carefully.

 This shows where you need to put your answer.

If you need to do any working out, you can use any space on the page.

Some questions have an answer box like this:

Show your working. You may get a mark.

For these questions you may get a mark for showing your working.

1   Circle the number closest in value to 6.

      0·6   7   6·5   5·6   5

1
‾‾‾‾‾
1 mark

2   Matt makes a fraction using two numbers.

He says 'My fraction is equivalent to $\frac{3}{4}$. One of my numbers is 12.'

Write the two fractions that he could have.

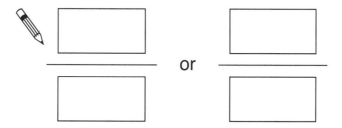

2
‾‾‾‾‾
2 marks

3   Guilio's Pizza Shop cuts each pizza into six slices.
They sell the slices for 95p per slice. A group of friends
buys 15 slices between them.

a) How many pizzas is this?

3a
‾‾‾‾‾
1 mark

b) How much do they pay altogether?

3b
‾‾‾‾‾
1 mark

4 Draw a line between each fraction and its equivalent decimal.

One has been done for you.

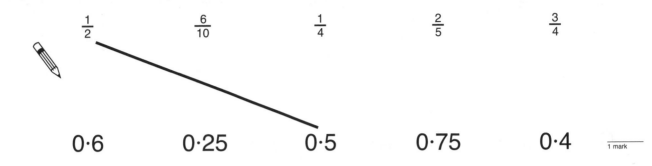

$\frac{1}{2}$ $\frac{6}{10}$ $\frac{1}{4}$ $\frac{2}{5}$ $\frac{3}{4}$

0·6 0·25 0·5 0·75 0·4 $\underset{\text{1 mark}}{}$ 4

5 Write the next two numbers in this sequence.

3　　12　　48　　192　　[ ]　　[ ] $\underset{\text{1 mark}}{}$ 5

6 Leah's mum is preparing party bags to give out at Leah's birthday party.

Each bag will have:

1 slice of cake

2 small toys

3 sweets

1 bag of crisps

Leah's mum packs 24 toys.

How many sweets does she pack?

Show your working. You may get a mark.

2 marks 6

7 Complete this number sentence.

 662 × ☐ = 5296

7

1 mark

8 265 people go to a concert.
Tickets costs £4.50 each.

a) How much ticket money is collected?

8a

1 mark

Programmes costs 80p each.
Selling programmes raises £75.20.

b) How many programmes are sold?

8b

1 mark

9 Write three different prime numbers that complete this
number sentence.

 ☐ × ☐ × ☐ = 2233

9

1 mark

10 Dan's watch showed this time when he left home.

When he reached school it showed this time.

How long did it take Dan to walk to school?

10

1 mark

11 Cindy bought $2\frac{1}{2}$ kg of cement and 6 identical bricks.

The total weight of her purchases was 7 kg 300 g.

How much did each brick weigh?

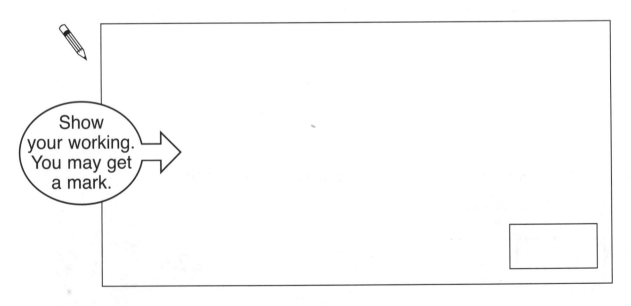

Show your working. You may get a mark.

11

2 marks

12 Here is a number sentence.

$$155 > \boxed{} \times 22$$

Circle all the numbers below that make the number sentence true.

   4   5   6   7   8   9

12
2 marks

13 Look at these digit cards.

| 2 | 4 | 6 | 7 |

Write all the three-digit numbers between 300 and 500 that can be made using these digit cards.

13
2 marks

14 Order these numbers from smallest to largest.

5·764   5·564   5·674   5·476   5·654

 [ ]   [ ]   [ ]   [ ]   [ ]

14
1 mark

smallest

15

A group of children did a sponsored walk. The graph below shows the results.

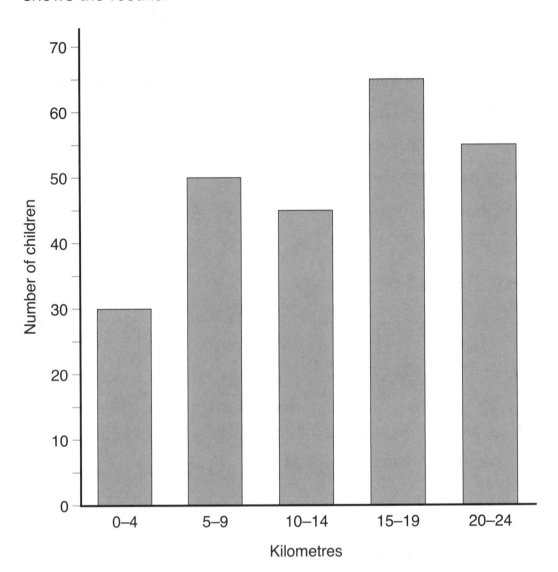

a) How many children walked a distance between 15 km and 19 km?

15a

1 mark

b) How many more children walked a distance between 5 km and 9 km than walked a distance between 0 km and 4 km?

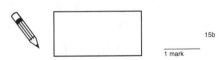

15b

1 mark

16 Draw a line from each calculation to the correct box.

One has been done for you.

$$5 \times 16 \times 5$$

multiple of 6

$$876 - 356$$

$$137 + 215 + 128$$

greater than 500

$$4545 \div 6$$

equal to 400

$$411 - (376 - 187)$$

16

2 marks

17 Each square is 1 cm². What is the area of this shape?

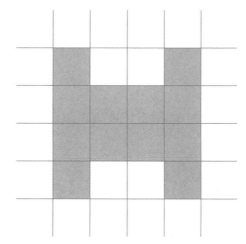

17

1 mark

18   Translate shape A 3 squares up and 4 squares to the left.

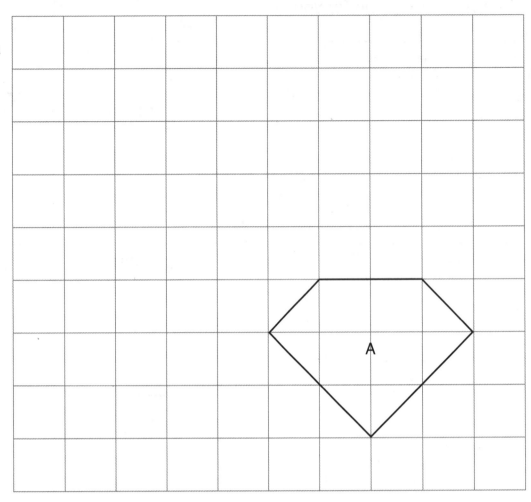

2 marks

19   Circle the weight that is nearest to the weight of an apple.

    2 g   20 g   100 g   800 g   2000 g

19

1 mark

20 Calculate the size of angle A.

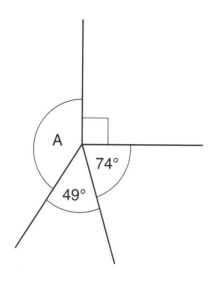

This diagram is not drawn accurately.

_____ 20

1 mark

21 Circle the nets below that are **not** the net of a closed cube.

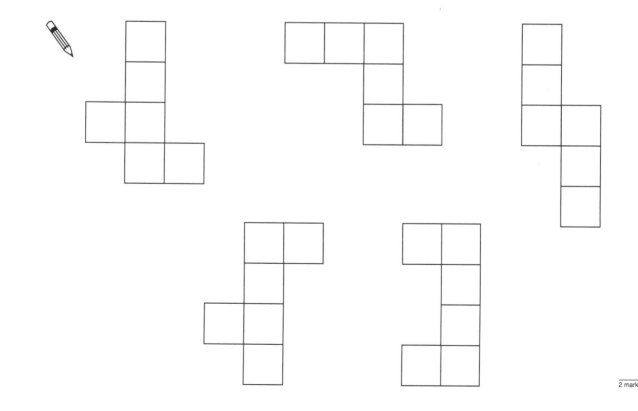

_____ 21

2 marks

22 Points A, B and C are three corners of a hexagon.

A line of symmetry runs from (6, 5) to (3, 2).

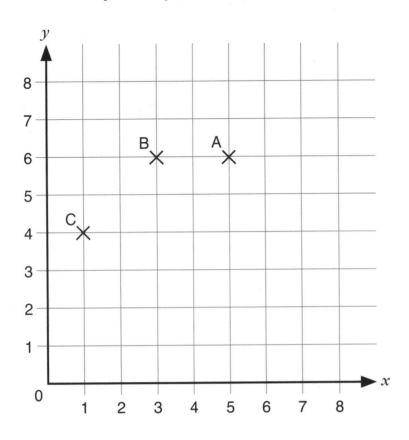

What are the co-ordinates of the other three corners?

22

3 marks

23 Sara buys 2·25 kg of apples.
Draw an arrow on the scale
to show this weight.

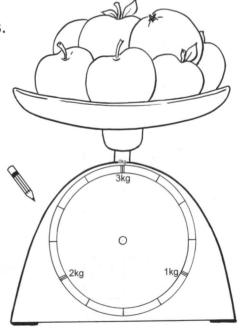

23

1 mark

Total out of 4 _____

24 100 people were asked about the different types of transport they used last week.

This is the result of the survey.

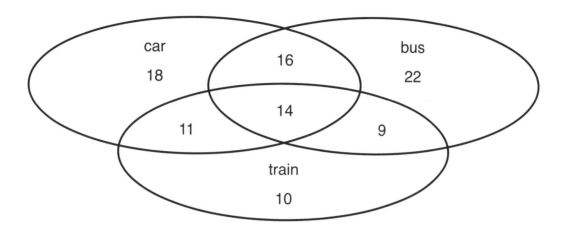

a) How many people travelled by train?

24a

1 mark

b) What type of transport did more people travel on?

24b

1 mark

25 Alex buys a pack of yoghurt for £3.96. There are 12 yoghurts in the pack, each containing 115 g of yoghurt.

a) How many kilograms of yoghurt does Alex buy?

25a

1 mark

b) How much does each pot of yoghurt cost?

25b

1 mark

End of test

Total out of 4 _____

Test 3

Mental mathematics test questions and answers

Say: **For this group of questions, you will have 5 seconds to work out each answer and write it down.**

	The questions	Answers
1.	What is double one point eight?	3·6
2.	What is three hundred and forty add sixty?	400
3.	How many eights are there in forty-eight?	6
4.	What is twenty out of eighty as a fraction in its simplest terms?	$\frac{1}{4}$
5.	Divide three hundred and sixty by ten.	36

Say: **For the next group of questions, you will have 10 seconds to work out each answer and write it down.**

		Answers
6.	What is the sum of one point three and two point four?	3·7
7.	Divide zero point one by one hundred.	0·001
8.	Add together one and a half, two and a half and four and a half.	$8\frac{1}{2}$
9.	What is four thousand, six hundred and thirty millilitres to the nearest litre?	5 litres
10.	Imagine a cylinder. How many faces does it have?	3
11.	Multiply zero point four by seven.	2·8
12.	Alpesh had one pound. He bought two ice lollies and got ten pence change. How much did each ice lolly cost?	45p
13.	The perimeter of a regular octagon is fifty-six centimetres. What is the length of each side?	7 cm
14.	How many minutes are there in five and a quarter hours?	315 min
15.	Seven times a number is four hundred and ninety. What is the number?	70

Say: **For the next group of questions, you will have 15 seconds to work out each answer and write it down.**

		Answers
16.	What is three sevenths of two hundred and ten?	90
17.	I think of a number, subtract five and halve the result. The answer is twenty-six. What is my number?	57
18.	Divide seventy-six by eight.	9·5
19.	In a bunch of thirty-six roses, there are twice as many red roses as white roses. How many white roses are there?	12
20.	Eight oranges cost three pounds twenty. How much do six oranges cost?	£2.40

Say: **Now put down your pencil. The test is finished.**

Test 3
Papers A and B answers

Paper A

1. 458, 485, 548, 584, 845, 854

2.
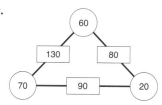

3. £385.23

4. 89

5. $\frac{12}{16} = \frac{3}{4} = \frac{9}{12} = \frac{15}{20}$

6. $x = 93°$ $y = 51°$

7. 5

8. $55\frac{1}{8}$ or 55·125

9.

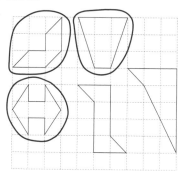

10.

	g	kg
Apple 1	220	**0·22**
Apple 2	**180**	0·18
Apple 3	**200**	0·2
Apple 4	195	**0·195**

11. £112

12. 260 ml

13.

14. 108

15. 9

16. 1268

17. 184 cm²

18. a) 13·1 cm   b) 25 mm

19.

	Number shot down	Points scored
Flying saucer	24	96
Rocket	48	144

20. a) ✗   b) ✓   c) ✓   d) ✗

21. a) 1 and 2

b) Yes, because there are 5 out of 10 sections on the spinner with number 4 on. $\frac{5}{10} = \frac{1}{2}$ which is an even chance.

22. a) 2 hr 30 min   b) 5 hr

23. a) 95 m   b) 11:30 a.m. and 12:00 p.m.

24. 36, 63, 81

25. a) 25%   b) 7

Paper B

1. 7, 1512

2. 6·37

3. Check that the children have written one correct number in each of the four sections of the sorting diagram.

4. $\frac{3}{4}$ or 0·75

5. 126 cm

6. 59 and 83

7.

5	4	7		5	4	9		5	8	9		5	8	7
	8	9			8	7			4	7			4	9

8. a) 75 ml   b) 925 ml

9. a) 19 km   b) 23·5 km

10. a) 75   b) Madrid ✈ ✈ ✈ ✈ ⊤

11. 22 °C

12. A = 53°   B = 37°

13.

TOTAL	Impossible	Poor chance	Even chance	Good chance	Certain
More than 3				✓	
Less than 4		✓			
0 or 1	✓				
Between 4 and 10				✓	
12 or less					✓

14.
regular hexagon	6
square	4
regular heptagon	7
rectangle	2

15. a) £140   b) £6.39

16. Accept 186 km or 198.4 km

17. 9

18. a) Because any child under 8 is included in the sub-sets 'Boys under 16' (7) and 'Girls under 16' (12).

b) 17

19. £314.10

20. a) $\frac{5}{6}$ or $\frac{11}{13}$ b) $\frac{1}{6}$ or $\frac{2}{13}$

21. 7

22. a) 7   b) 6

23. a) 15   b) 30

24.

Mental mathematics test

Name ....

Date .... Class ....

Total marks

Time: 5 seconds

1		1·8	___ 1
2		340 60	___ 2
3		48	___ 3
4		20 80	___ 4
5		360	___ 5

11		0·4	___ 11
12	p	£1 10p	___ 12
13		56 cm	___ 13
14	min	$5\frac{1}{4}$ hr	___ 14
15		490	___ 15

Time: 10 seconds

6		1·3 2·4	___ 6
7		0·1	___ 7
8		$1\frac{1}{2}$ $2\frac{1}{2}$ $4\frac{1}{2}$	___ 8
9	litres	4630 ml	___ 9
10			___ 10

Time: 15 seconds

16		210	___ 16
17		5 26	___ 17
18		76 8	___ 18
19		36	___ 19
20	£	£3.20	___ 20

Year 6 Test 3

MATHEMATICS

YEAR 6

TEST 3 PAPER A

LEVELS 3–5

CALCULATOR NOT ALLOWED

PAGE	MARKS
3	
4	
5	
6	
7	
8	
9	
10	
11	
12	
13	
TOTAL	

RESOURCES
- pencil

Name

Date

Class

Instructions

You may not use a calculator to answer any questions in this test.

Work as quickly and as carefully as you can.

You have 45 minutes for this test.

If you cannot do one of the questions, go on to the next one.

You can come back to it later if you have time.

If you have finished before the end, go back and check your work.

Follow the instructions for each question carefully.

 This shows where you need to put your answer.

If you need to do any working out, you can use any space on the page.

Some questions have an answer box like this:

For these questions you may get a mark for showing your working.

1 Arrange the following numbers in order, smallest to largest.

548   854   458   584   485   845

smallest

2 Write these numbers in the correct places in the diagram.

4   7   8   9

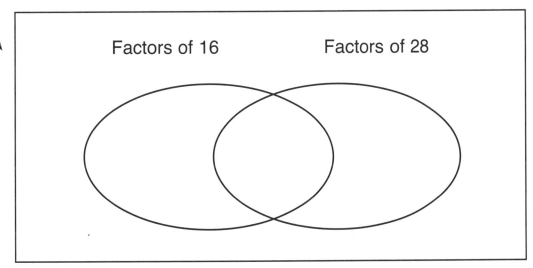

3 Calculate £76.46 + £308.77

4 Complete the following sequence.

73   81   [   ]   97   105

1 mark

5 Complete these fractions so that they are equivalent.

$$\frac{\boxed{}}{16} = \frac{3}{4} = \frac{\boxed{}}{12} = \frac{\boxed{}}{20}$$

5

2 marks

6 Look at the triangle below.

Calculate the size of angles x and y.

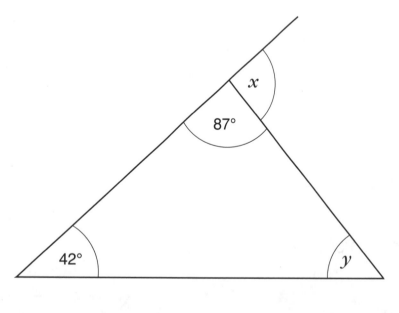

This diagram is not to scale.

87°

x

42°

y

x [   ] y [   ]

6

2 marks

7 The chickens on a farm lay 114 eggs a day. The eggs are packed in trays holding 24 eggs each.

How many trays are needed each day?

7

1 mark

8 Calculate 882 ÷ 16

Show your working. You may get a mark.

8

2 marks

9 Fill in the missing numbers in the box and circles so that the number in each box is the sum of the numbers in the circles joined to it.

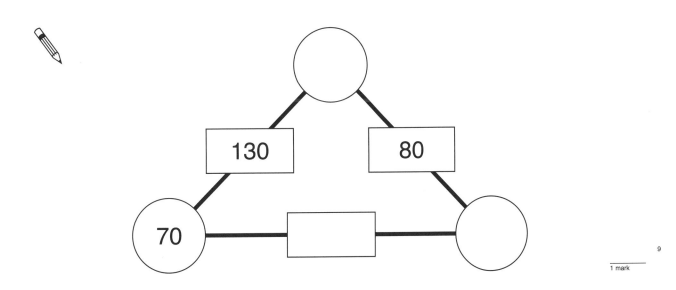

130

80

70

9

1 mark

10 This table shows the weight of four apples.

Complete the table.

	g	kg
Apple 1	220	
Apple 2		0·18
Apple 3		0·2
Apple 4	195	

10

2 marks

11 Calculate 35% of £320

11

1 mark

12 How much liquid is in this jug?

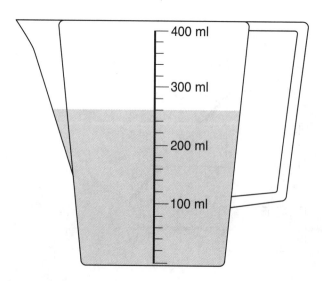

12

1 mark

13 Here are five shapes on a grid.

Draw a circle around the 3 shapes that have reflective symmetry.

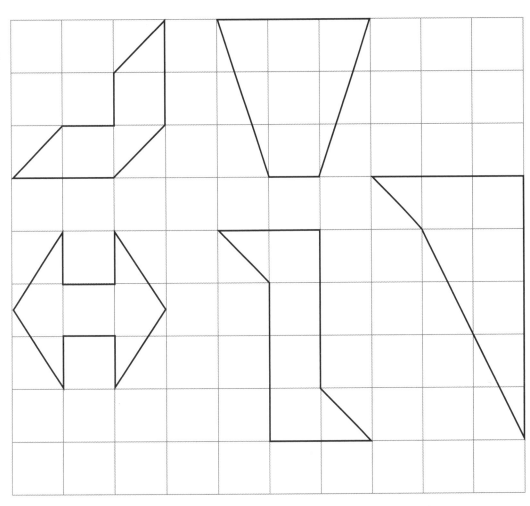

3 marks

14 Calculate $\frac{4}{9} \times 243$

14

1 mark

Total out of 4 _____

15 Write the missing number in the box.

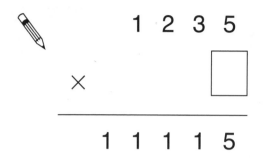

$$\begin{array}{r} 1\ 2\ 3\ 5 \\ \times \qquad \boxed{} \\ \hline 1\ 1\ 1\ 1\ 5 \end{array}$$

1 mark

16 Calculate 1506 − 238

16

1 mark

17 What is the area of this shape?

This diagram is not to scale.

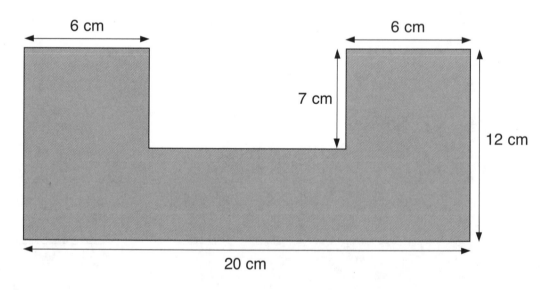

17

2 marks

18 Here is a calculator, a 10p coin and an eraser.

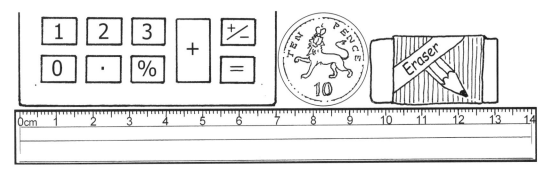

a) What is the length of all three things together?
Give your answer in centimetres.

18a

1 mark

b) Look at the ruler above.
What is the diameter of the 10p coin?
Give your answer in millimetres.

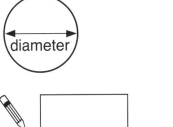

18b

1 mark

19 Sim is playing *Space Attacks* on his computer.
He scores 4 points for each flying saucer he
shoots down and 3 points for each rocket.

4 points

He scores 96 points by shooting down flying
saucers. He shoots down twice as many
rockets as flying saucers.

3 points

Complete the table below.

	Number shot down	Points scored
		96

19

2 marks

20 Look at this shape.

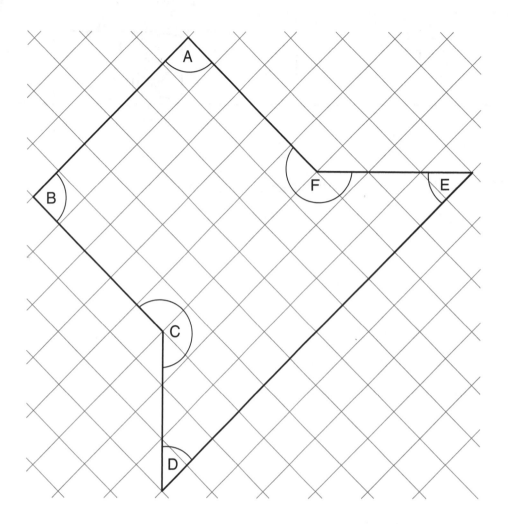

For each statement, put a tick (✓) if it is true or a cross (✗) if it is false.

a) Angle A is an acute angle.

b) Angle C is an obtuse angle.

c) Side AB is parallel to side DE.

d) Side BC is perpendicular to side CD.

20

2 marks

21 This spinner is divided into 10 equal sections.

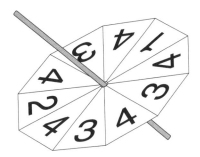

a) Which two different numbers on the spinner are equally likely to come up?

and

b) Brian says *'4 has got an even chance of coming up.'*
   Is Brian correct? Explain why?

22 This table shows Jon's journey to his aunt's house.

8:30 a.m.	Left home
10:00 a.m.	Got on train
12:30 p.m.	Got off train
1:30 p.m.	Arrived at aunt's house

a) How long was Jon's train ride?

b) How long after leaving home did Jon arrive at his aunt's house?

23 Jane and Gita climb a mountain. The line graph below shows how high they are as they climb to the top.

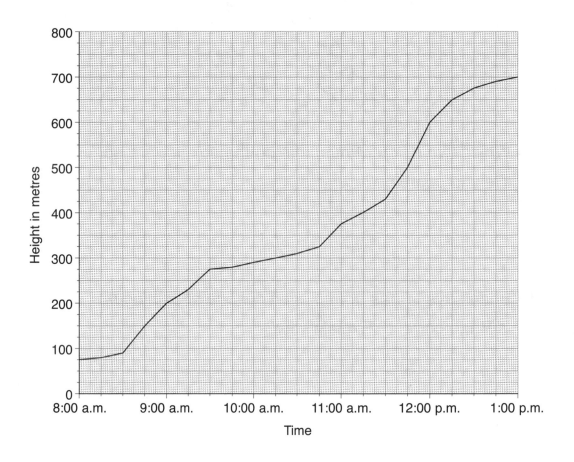

a) What is their change in height between 9:15 a.m. and 10:45 a.m.?

23a

1 mark

b) Over which half-hour period is their climb the steepest?

23b

1 mark

24 Circle all the multiples of 9.

26   36   53   63   71   81

24

1 mark

25 A class was asked what their favourite drink was.

The pie chart below shows the results.

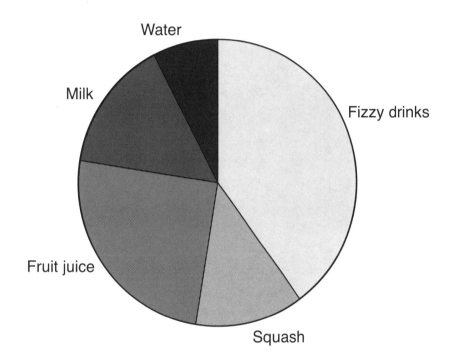

a) What percentage of the class preferred fruit juice?

25a

1 mark

14 children preferred fruit juice.

b) How many children preferred squash?

25b

1 mark

End of test

MATHEMATICS

YEAR 6

TEST 3 PAPER B   LEVELS 3–5

CALCULATOR ALLOWED

PAGE	MARKS
3	
4	
5	
6	
7	
8	
9	
10	
11	
12	
13	
TOTAL	

RESOURCES

- pencil
- calculator

Name

Date

Class

Instructions

You may use a calculator to answer any questions in this test.

Work as quickly and as carefully as you can.

You have 45 minutes for this test.

If you cannot do one of the questions, go on to the next one.

You can come back to it later if you have time.

If you have finished before the end, go back and check your work.

Follow the instructions for each question carefully.

 This shows where you need to put your answer.

If you need to do any working out, you can use any space on the page.

Some questions have an answer box like this:

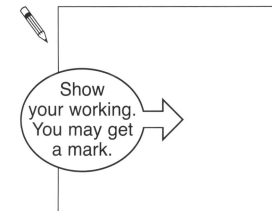

For these questions you may get a mark for showing your working.

1 Junior made a sequence of numbers.

His rule was:

Double the previous number, then multiply by 3

Write in the missing numbers in his sequence.

 ☐ 42 252 ☐ 9072

2 marks

2 Circle the number that is closest to 6·4

 6·48 6·5 6 6·37 6·45

1 mark

3 Look at the sorting diagram.

Write one number in each section of the diagram.

	Less than 300	More than 300
Even number divisible by 3		
Odd number divisible by 3		

2 marks

4 Here is a number line.

Estimate the number marked by the arrow.

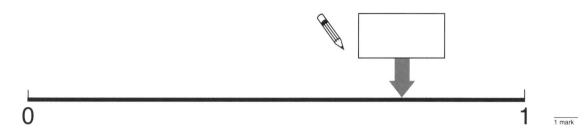

0   1

1 mark

5 This shape is made from identical squares.

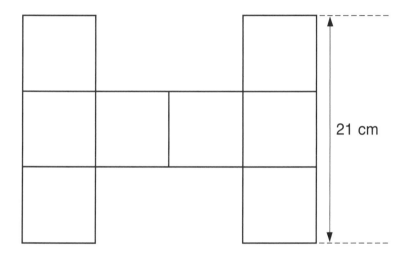

21 cm

What is its perimeter?

5

2 marks

6 Circle the two prime numbers.

     49   59   63   83   99

6

1 mark

7 Look at the six digit cards below.

| 4 | 5 | 6 | 7 | 8 | 9 |

Use five of the digit cards to make the following calculation correct.

$$\begin{array}{r} \square\ \square\ \square \\ +\quad\square\ \square \\ \hline 6\ \ 3\ \ 6 \\ \hline \end{array}$$

7

2 marks

8 Julie is making a chocolate milkshake for herself and a friend.

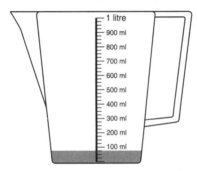

a) How much chocolate sauce has Julie used?

8a

1 mark

Julie then adds 850 ml of milk.

b) How much chocolate milkshake does Julie make?

8b

1 mark

Total out of 4 _____

9 Rebecca has just walked 12·5 km from Bomaderry.

a) If Rebecca walks from Bomaderry to Nowra, how far will she have walked?

9a

1 mark

b) What is the distance from Nowra to Berry?

9b

1 mark

10 This pictogram shows where the people waiting in an airport lounge are travelling.

Paris	✈✈✈✈✈✈✈⊤
Rome	✈✈✈
Madrid	
Berlin	✈✈
Athens	✈✈✈✈✈✈✈⊤

KEY

✈ = 10 people

a) How many people are travelling to Paris?

b) 45 people are travelling to Madrid. Show this on the pictogram.

11 The temperature in a fridge is 4 °C. The temperature in the freezer compartment is −18 °C.

What is the difference in temperature between the two?

12 Look at the triangle below.

Calculate the size of angles A and B.

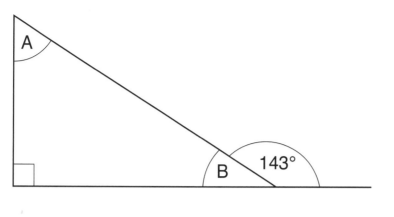

This diagram is not to scale.

A ☐   B ✏ ☐

12

2 marks

13 Tan is throwing two 1–6 dice and adding the numbers together.
What are his chances of getting the following totals?

One has been done for you.

TOTAL	Impossible	Poor chance	Even chance	Good chance	Certain
More than 3				✓	
Less than 4					
0 or 1					
Between 4 and 10					
12 or less					

13

2 marks

14 How many lines of symmetry do the following shapes have?

regular hexagon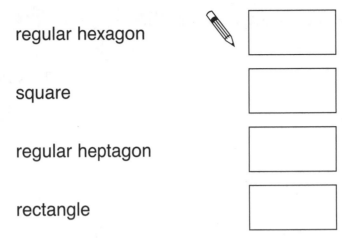

square

regular heptagon

rectangle

14

15 Jensen gets an electricity bill for £34.67, a gas bill for £57.80 and a telephone bill for £53.92.

a) He rounds each bill to the nearest £10 and adds them together. How much is this?

15a

1 mark

b) He later adds them up accurately. What is the difference between the rounded total and the accurate total?

15b

1 mark

16 The distance between London and Bristol is 124 miles.

Approximately how far is this in kilometres?

16

1 mark

17 45 children went on a school trip. 60% were girls. $\frac{1}{3}$ of the girls had blue eyes.

How many girls with blue eyes went on the trip?

Show your working. You may get a mark.

17

2 marks

18 This Venn diagram shows the people under 16 years old who live in Baker Street.

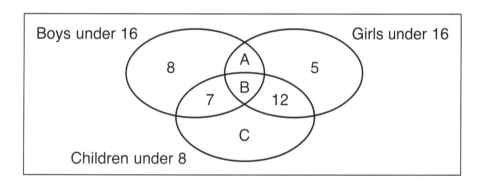

Boys under 16   Girls under 16

8   A   5

B

7   12

C

Children under 8

a) Sections A and B will always be empty because no child can be in both the boy and girl sections. Describe why section C will also always be empty.

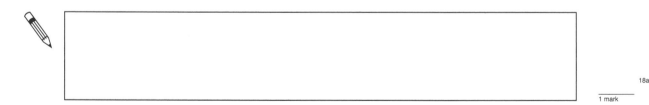

18a

1 mark

b) How many girls under 16 live in Baker Street?

18b

1 mark

19 Calculate 45% of £698

19

1 mark

20

a) Use two of these number cards to make a fraction greater than $\frac{3}{4}$.

20a

1 mark

b) How much less than 1 is your fraction?

20b

1 mark

21 $3x + 12 = 33$

What is the value of x?

21

1 mark

22 a) Anna thinks of a number. She squares it and then
    subtracts 5. Her answer is 44. What was her number?

22a
_____
1 mark

b) Leroy thinks of a number. He adds 32 to the number
   and then halves the total. His answer is 19. What was
   his number?

22b
_____
1 mark

23 Zina was counting the birds she saw in the park one afternoon.
    She counted 45 birds in total.

One in every three birds was a sparrow.

a) How many sparrows did she count?

23a
_____
1 mark

b) How many of the birds she counted were not sparrows?

23b
_____
1 mark

24 Draw lines to match the triangles to their names.

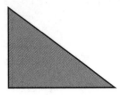  

| isosceles | | scalene | | equilateral |

24

1 mark

End of test

Total out of 1 _____

Record-keeping format 1 Adult Directed Task assessment sheet

Objective(s): _____

Adult: _____

Date: _____

Class: _____

NC Level: _____

Child's name	Success criteria				Other observations	Objective(s) achieved	Future action

Record-keeping format 2 Test 1 grid for test analysis (Mental mathematics test)

	AT	Question	Mark
Addition: multiples of 10	2	1	1
Multiplication tables	2	2	1
Subtraction: 100 less	2	3	1
2-D shape	3	4	1
Rounding decimals	2	5	1
Calculating positive and negative numbers	2	6	1
Addition: multiples of 10	2	7	1
Addition	2	8	1
Addition: decimals	2	9	1
Fractions/decimals	2	10	1
3-D solids	3	11	1
Length: multiplication	2 & 3	12	1
Subtraction: decimals	2	13	1
Addition: fractions	2	14	1
2-D shape/perimeter	3	15	1
Multiplication: money	1 & 2	16	1
Multiplication: multiples of 10	2	17	1
Addition/Multiplication	2	18	1
Subtraction	2	19	1
Multiplication/addition: money	1 & 2	20	1

Mental mathematics test score (out of 20)

Name: 1. 2. 3. 4. 5. 6. 7. 8. 9. 10. 11. 12. 13. 14. 15. 16. 17. 18. 19. 20. 21. 22. 23. 24. 25. 26. 27. 28. 29. 30.

Number correct
Number incorrect or omitted
Percentage correct
Percentage incorrect or omitted

Record-keeping format 2   Test 1 grid for test analysis (Paper A)

Topic	Strand	Q	Marks	1.	2.	3.	4.	5.	6.	7.	8.	9.	10.	11.	12.	13.	14.	15.	16.	17.	18.	19.	20.	21.	22.	23.	24.	25.	26.	27.	28.	29.	30.
Ordering whole numbers	2	1	1																														
Sequences	2	2	1																														
Divisibility	2	3	1																														
Handling data: Venn diagram / factors	2 & 4	4	2																														
Subtraction	2	5	1																														
Multiplication (inverse)	2	6	1																														
2-D shape	3	7	2																														
Reflection	3	8	2																														
Fractions/decimals	2	9	1																														
Multiplication	1 & 2	10	2																														
Ratio	1 & 2	11ab	2																														
Area	3	12	2																														
Multiplication	1 & 2	13	1																														
Percentages of shapes	2	14	1																														
Time	3	15	1																														
Division	1 & 2	16	1																														
Long multiplication	2	17	2																														
Rounding decimals	2	18	1																														
a. Multiplication/addition b. Subtraction	1 & 2	19ab	2																														
Perimeter	3	20	2																														
Handling data: pie chart	4	21ab	2																														
1st quadrant co-ordinates	2 & 3	22ab	2																														
Calculating positive and negative numbers	2	23	1																														
Handling data: bar chart	4	24ab	2																														
Handling data: table	4	25ab	2																														
Handling data: line graph	4	26ab	2																														

Paper A score (out of 40)

Record-keeping format 2 Test 1 grid for test analysis (Paper B)

Question/Topic	NC Level	Q no.	Marks
Sequences	2	1	1
a. Subtraction (inverse) b. Division (inverse) c. Multiplication (inverse)	2	2abc	3
Ordering decimals	2	3	1
Division, multiplication and subtraction: rounding	2	4	2
Calculating (inverse)	1&2	5ab	2
Sequences / square numbers	2	6	1
Ordering positive and negative integers	2	7	1
Percentages of amounts	2	8	1
Handling data: Timetables	4	9ab	2
Length/time	1&3	10	2
Mass: scales	3	11	1
Algebra	2	12	2
Time	3	13	1
a. Multiplication: money b. Division: money	1&2	14ab	2
a. Multiplication (inverse) b. Subtraction (inverse)	2	15ab	2
Angles in a triangle	3	16AB	2
Subtraction, division and addition	1&2	17	1
Sequence: subtraction	2	18	1
a. Addition and subtraction: money b. Division	1&2	19ab	2
Multiplication and addition: mass	1&3	20	2
Fractions of amounts	2	21	1
Handling data: 2-D shape	3&4	22	2
Probability scale	4	23	1
a. Capacity: scales b. Subtraction: capacity	1,2&3	24ab	2
Reflective symmetry	3	25	2

Summary columns:
- Paper B score (out of 40)
- Total score (out of 100)
- National Curriculum Level

Student rows: 1.–30. (blank)

Record-keeping format 3   Test 2 grid for test analysis (Mental mathematics test)

Mental mathematics test score (out of 20)

AT	Question	Mark	Topic
2	1	1	Fractions/division
2	2	1	Writing numbers
3	3	1	Time
2	4	1	Subtraction
3	5	1	Multiplication: mass
2	6	1	Fractions: money
3	7	1	2-D shape: area
2	8	1	Addition
2	9	1	Decimals/percentages
2	10	1	Multiplication
2	11	1	Multiplication: decimals
2	12	1	Fractions of amounts
2	13	1	Multiplication
3	14	1	3-D solids
2	15	1	Percentages of amounts
2	16	1	Subtraction: decimals
2	17	1	Multiplication
3	18	1	Calculating angles
2	19	1	Division: decimals
1 & 2	20	1	Percentages

Name: 1. – 30.

Number correct

Number incorrect or omitted

Percentage correct

Percentage incorrect or omitted

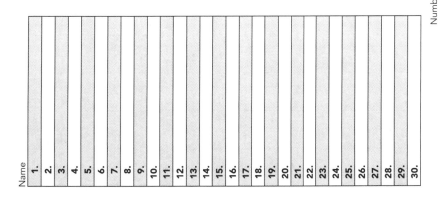

173

Record-keeping format 3 Test 2 grid for test analysis (Paper A)

	Ordering whole numbers	Ordering whole numbers	Multiplication (inverse)	Square numbers	Calculating positive and negative integers	Sequences	Division	Properties of numbers	Temperature difference	Rounding decimals	Handling data: line graph	Multiplication: length	Subtraction	Multiplication, addition and subtraction	Area	Handling data: Venn diagram / factors	Percentages	Handling data: pie chart	Ratio and proportion	Perimeter	Divisibility	Handling data: averages	Time	Angles	Rotation	Paper A score (out of 40)
	2	2	2	2	1&2	2	2	1,2&4	2	1&2	4	1,2&3	2	1&2	3	2&4	1&2	4	1&2	3	2	4	1&2	3	3	
	1	2	3	4	5	6	7	8ab	9	10	11ab	12	13	14	15	16	17	18ab	19	20	21	22ab	23	24AB	25	
	2	1	1	1	2	1	2	2	1	1	2	1	1	2	2	2	1	2	2	2	1	2	2	2	2	
1.																										
2.																										
3.																										
4.																										
5.																										
6.																										
7.																										
8.																										
9.																										
10.																										
11.																										
12.																										
13.																										
14.																										
15.																										
16.																										
17.																										
18.																										
19.																										
20.																										
21.																										
22.																										
23.																										
24.																										
25.																										
26.																										
27.																										
28.																										
29.																										
30.																										

Record-keeping format 3 Test 2 grid for test analysis (Paper B)

Topic	NC level	Q no.	Marks
Rounding decimals	2	1	1
Equivalent fractions	1&2	2	2
a. Fractions / b. Multiplication	1&2	3ab	2
Fractions and decimals	2	4	1
Sequences	2	5	1
Ratio and Proportion	1&2	6	2
Multiplication (inverse)	2	7	1
a. Multiplication / b. Division	1&2	8ab	2
Prime numbers / multiplication	1&2	9	1
Time	3	10	1
Subtraction and division: mass	1,2&3	11	2
Multiplication/ Inequalities	2	12	2
Properties of numbers	1&2	13	2
Ordering decimals	2	14	1
Handling data: bar chart	4	15ab	2
Approximation / calculating	2	16	2
Area	3	17	1
Translation	3	18	2
Mass: scales	3	19	1
Angle	3	20	1
3-D shape: nets	3	21	2
2-D shape / co-ordinates	2&3	22	3
Mass	3	23	1
Handling data: Venn diagram	4	24ab	2
a. Multiplication: mass / b. Division	1,2&3	25ab	2

Summary columns:
- Paper B score (out of 40)
- Total score (out of 100)
- National Curriculum Level

Pupil rows: 1. 2. 3. 4. 5. 6. 7. 8. 9. 10. 11. 12. 13. 14. 15. 16. 17. 18. 19. 20. 21. 22. 23. 24. 25. 26. 27. 28. 29. 30.

Record-keeping format 4 Test 3 grid for test analysis (Mental mathematics test)

Question topic	AT	Question	Mark
Multiplication: decimals	2	1	1
Addition	2	2	1
Division	2	3	1
Reducing fractions	2	4	1
Division	2	5	1
Addition: decimals	2	6	1
Division: decimals	2	7	1
Addition: fractions	2	8	1
Capacity: rounding	3	9	1
3-D solids	3	10	1
Multiplication: decimals	2	11	1
Subtraction/division (inverse)	1 & 2	12	1
2-D shape: perimeter	3	13	1
Multiplication: time	3	14	1
Multiplication (inverse)	2	15	1
Fractions of amounts	2	16	1
Subtraction/division (inverse)	1 & 2	17	1
Division	2	18	1
Ratio and proportion	1 & 2	19	1
Direct proportion	1 & 2	20	1

Mental mathematics test score (out of 20)

Name: 1. 2. 3. 4. 5. 6. 7. 8. 9. 10. 11. 12. 13. 14. 15. 16. 17. 18. 19. 20. 21. 22. 23. 24. 25. 26. 27. 28. 29. 30.

Number correct

Number incorrect or omitted

Percentage correct

Percentage incorrect or omitted

Record-keeping format 4 Test 3 grid for test analysis (Paper A)

	Ordering whole numbers	Handling data: Venn diagram / factors	Addition: money	Sequences	Equivalent fractions	Angles in a triangle	Division: rounding	Division	Addition (inverse)	Multiplication and division: mass	Percentages of amounts	Capacity: scales	2-D shapes: symmetry	Fractions of amounts	Multiplication (inverse)	Subtraction	Area	Length	Multiplication and division	Angles and lines	Probability	Handling data: tables	Handling data: line graph	Multiples	Handling data: pie chart
	2	2&4	2	2	2	3	1&2	2	1&2	2&3	2	3	3	2	2	2	3	3	1&2	3	4	4	4	2	4
	1	2	3	4	5	6xy	7	8	9	10	11	12	13	14	15	16	17	18ab	19	20	21ab	22ab	23ab	24	25ab
	1	2	1	1	2	2	1	2	1	2	1	1	3	1	1	1	2	2	2	2	2	2	2	1	2
1.																									
2.																									
3.																									
4.																									
5.																									
6.																									
7.																									
8.																									
9.																									
10.																									
11.																									
12.																									
13.																									
14.																									
15.																									
16.																									
17.																									
18.																									
19.																									
20.																									
21.																									
22.																									
23.																									
24.																									
25.																									
26.																									
27.																									
28.																									
29.																									
30.																									

Paper A score (out of 40)

Record-keeping format 4 Test 3 grid for test analysis (Paper B)

Topic	Level	Q	Marks
Sequencing	1&2	1	2
Rounding decimals	2	2	1
Handling data: table / properties of numbers	2&4	3	2
Decimals	2	4	1
Perimeter	3	5	2
Prime numbers	2	6	1
Addition (inverse)	1&2	7	2
Capacity: scales	3	8ab	2
Addition: length/distance	1,2&3	9ab	2
Handling data: pictograms	4	10ab	2
Subtracting positive and negative integers	2	11	1
Angles	3	12AB	2
Handling data: table / probability	4	13	2
2-D shape/symmetry	3	14	2
a. Rounding/addition b. Addition/subtraction	1&2	15ab	2
Length: metric/imperial conversion	2&3	16	1
Percentages and fractions	2	17	2
Handling data: Venn diagram	1&4	18ab	2
Percentages of amounts	2	19	1
Fractions	1&2	20ab	2
Algebra	1&2	21	1
4 operations (inverse)	1&2	22ab	2
Proportion	2	23ab	2
2-D shapes: triangles	3	24	1

Additional score columns: Paper B score (out of 40); Total score (out of 100); National Curriculum Level.

Student rows: 1. to 30. (blank recording grid)

Record-keeping format 5 Attainment Target 1 – Using and applying mathematics

Level 2

Problem solving
- Select and use material in some classroom activities
- Select and use mathematics for some classroom activities
- Begin to develop own strategies for solving a problem
- Begin to understand ways of working through a problem

Communicating
- Discuss work using mathematical language
- Respond to and ask mathematical questions
- Begin to represent work using symbols and simple diagrams
- Explain why an answer is correct

Reasoning
- Ask questions such as: 'What would happen if...?' 'Why?'
- Begin to develop simple strategies

Level 3

Problem solving
- Develop different mathematical approaches to a problem
- Look for ways to overcome difficulties
- Begin to make decisions and realise that results may vary according to the 'rule' used
- Begin to organise work
- Check results

Communicating
- Discuss mathematical work
- Begin to explain thinking
- Use and interpret mathematical symbols and diagrams

Reasoning
- Understand a general statement
- Investigate general statements and predictions by finding and trying out examples

Level 4

Problem solving
- Develop own strategies for solving problems
- Use own strategies for working within mathematics
- Use own strategies for applying mathematics to practical contexts

Communicating
- Present information and results in a clear and organised way

Reasoning
- Search for solutions by trying out own ideas

Level 5

Problem solving
- Identify and obtain necessary information
- Check results, considering whether these are sensible

Communicating
- Show understanding of situations by describing them mathematically using symbols, words and diagrams

Reasoning
- Draw simple conclusions
- Give an explanation for their reasoning

General comments

179

Record-keeping format 6   Attainment Target 2 – Number and algebra

Level 2

Numbers and the number system	**Calculations**	**Solving numerical problems**
• Count sets of objects reliably • Understand place value (HTU) • Order numbers up to 100 • Recognise sequences of numbers • Recognise odd and even numbers	• Recall addition and subtraction number facts to 10 • Understand that subtraction is the inverse of addition	• Use appropriate operation • Use mental strategies to solve problems involving money and measures

Level 3

Numbers and the number system	**Calculations**	**Solving numerical problems**
• Understand place value (ThHTU) • Begin to use decimal notation • Recognise negative numbers • Use simple fractions that are several parts of a whole • Recognise when two fractions are equivalent	• Make approximations • Recall addition and subtraction number facts to 20 • Add and subtract two two-digit numbers mentally • Add and subtract three two-digit numbers using written methods • Recall 2, 3, 4, 5, 10 multiplication tables • Recall division facts corresponding to the 2, 3, 4, 5, 10 multiplication tables	• Solve word problems involving larger numbers • Solve word problems involving multiplication • Solve word problems involving division, including those with a remainder

Level 4

Numbers and the number system	**Calculations**	**Solving numerical problems**
• Multiply and divide whole numbers by 10 or 100 • Add and subtract decimals to two places • Order decimals to three places • Recognise approximate proportions of a whole • Use simple fractions and percentages to describe proportions of a whole • Recognise and describe number patterns • Recognise and describe a multiple, factor and square	• Recall multiplication facts up to 10 × 10 • Recall division facts corresponding to the multiplication facts up to 10 × 10 • Use efficient written methods for addition and subtraction • Use efficient written methods for short multiplication and division	• Use a range of mental calculation strategies for the four operations • Check the reasonableness of an answer • Begin to use simple formulae expressed in words • Use and interpret co-ordinates in the first quadrant

Level 5

Numbers and the number system	**Calculations**	**Solving numerical problems**
• Multiply and divide whole numbers and decimals by 10, 100 and 1000 • Order negative numbers • Add and subtract negative numbers • Reduce a fraction to its simplest form • Solve simple problems involving ratio and proportion • Calculate fractional or percentage parts of quantities and measurements	• Use brackets appropriately • Use efficient written methods for addition and subtraction up to 10 000 • Use efficient written methods for long multiplication and division • Use all four operations with decimals to two places	• Check solutions by applying inverse operations • Check solutions by estimating using approximations • Construct and express in symbolic form simple formulae involving one or two operations • Use and interpret co-ordinates in all four quadrants

General comments

Year 6 National Expectations
Start-of-year: Level 3a (4c)
End-of-year: Level 4b

Record-keeping format 7 Attainment Target 3 – Shape, space and measures

Level 2

Understanding properties of shapes	Understanding properties of position and movement	Understanding measures	
• Use mathematical names for common 2-D and 3-D shapes • Describe the properties of common 2-D and 3-D shapes, including number of sides and corners	• Distinguish between straight and turning movements • Understand angle as a measure of turn • Recognise right angles in turns	• Begin to use everyday non-standard units to measure length and mass • Begin to use everyday standard units to measure length and mass	

Level 3

Understanding properties of shapes	Understanding properties of position and movement	Understanding measures	
• Classify 3-D and 2-D shapes in various ways	• Use mathematical properties such as reflective symmetry to describe 2-D shapes	• Use non-standard units of length, capacity and mass in a range of contexts • Use standard units of length, capacity, mass and time in a range of contexts	

Level 4

Understanding properties of shapes	Understanding properties of position and movement	Understanding measures	
• Make 3-D mathematical models by linking given faces or edges • Draw common 2-D shapes in different orientations on grids	• Reflect simple shapes in a mirror line	• Choose and use appropriate units • Choose and use appropriate instruments • Interpret, with appropriate accuracy, numbers on a range of measuring instruments • Find perimeters of simple shapes • Find areas by counting squares	

Level 5

Understanding properties of shapes	Understanding properties of position and movement	Understanding measures	
• Measure and draw angles to the nearest degree when constructing models and drawing shapes	• Use language associated with angle • Know the angle sum of a triangle • Know the sum of angles at a point • Identify all the symmetries of 2-D shapes	• Know the rough metric equivalences of imperial units still in daily use • Convert one metric unit to another • Make sensible estimates of a range of measures in relation to everyday situations • Understand and use the formula for the area of a rectangle	

General comments

Year 6 National Expectations
Start-of-year: Level 3a (4c)
End-of-year: Level 4b

Record-keeping format 8　Attainment Target 4 – Handling data

Level 2

Processing, representing and interpreting data
- Sort objects using more than one criterion
- Classify objects using more than one criterion
- Record results in simple lists, tables and block graphs
- Communicate findings

Level 3

Processing, representing and interpreting data
- Extract and interpret information presented in simple tables and lists
- Construct and interpret bar charts and pictograms where the symbol represents a group of units

Level 4

Processing, representing and interpreting data
- Collect discrete data and record them using a frequency table
- Understand and use the mode and range to describe sets of data
- Group data, where appropriate, in equal class intervals
- Represent collected data in frequency diagrams and interpret such diagrams
- Construct and interpret simple line graphs

Level 5

Processing, representing and interpreting data
- Understand and use the mean of discrete data
- Compare two simple distributions using the range and one of the mode, median or mean
- Interpret graphs and diagrams, including pie charts, and draw conclusions
- Understand and use the probability scale from 0 to 1
- Find and justify probabilities
- Select and use methods based on equally likely outcomes and experimental evidence
- Understand that different outcomes may result from repeating an experiment

General comments

Year 6 National Expectations
Start-of-year: Level 3a (4c)
End-of-year: Level 4b

182

Record-keeping format 9 Class record of the end-of-year expectations

Class: _____

Date: _____

Names

Year 6 End-of-year expectations																							
Counting and understanding number (AT2) Express one quantity as a percentage of another; find equivalent percentages, decimals and fractions (Level 4)																							
Knowing and using number facts (AT2) Use knowledge of place value and multiplication facts to 10 × 10 to derive related multiplication and division facts involving decimals (Level 4)																							
Calculating (AT2) Use efficient written methods to add and subtract integers and decimals, to multiply and divide integers and decimals by a one-digit integer, and to multiply two- and three-digit integers by a two-digit integer (Level 4)																							
Understanding shape (AT3) Visualise and draw on grids of different types where a shape will be after reflection, after translations, or after rotation through 90° or 180° about its centre or one of its vertices (Level 4)																							
Measuring (AT3) Select and use standard metric units of measure and convert between units using decimals to two places (Level 4)																							
Handling data (AT4) Solve problems by collecting, selecting, processing, presenting and interpreting data, using ICT where appropriate; draw conclusions and identify further questions to ask (Level 4)																							

NOTES: **Using and applying mathematics (AT1)** is incorporated throughout
End-of-year National Expectations: Level 4b

Record-keeping format 10   Individual child's record of the end-of-year expectations

Name: _____

Foundation Stage	Year 1	Year 2	Year 3
Using and applying mathematics (AT1)			
Use developing mathematical ideas and methods to solve practical problems (Level 1)			
Talk about, recognise and recreate simple patterns (Level 1)			
Counting and understanding number (AT2)			
Say and use the number names in order in familiar contexts (Level 1)	Read and write numerals from 0 to 20, then beyond; use knowledge of place value to position these numbers on a number track and number line (Level 2)	Count up to 100 objects by grouping them and counting in tens, fives or twos; explain what each digit in a two-digit number represents, including numbers where 0 is a place holder; partition two-digit numbers in different ways, including into multiples of ten and one (Level 2)	Partition three-digit numbers into multiples of one hundred, ten and one in different ways (Level 2)
Count reliably up to 10 everyday objects (Level 1)			
Use language such as 'more' or 'less' to compare two numbers (Level 1)			
Recognise numerals 1 to 9 (Level 1)			
Knowing and using number facts (AT2)			
Find one more or one less than a number from 1 to 10 (Level 1)	Derive and recall all pairs of numbers with a total of 10 and addition facts for totals to at least 5; work out the corresponding subtraction facts (Level 2)	Derive and recall all addition and subtraction facts for each number to at least 10, all pairs with totals to 20 and all pairs of multiples of 10 with totals up to 100 (Level 2)	Derive and recall all addition and subtraction facts for each number to 20, sums and differences of multiples of 10 and number pairs that total 100 (Level 3)
Calculating (AT2)			
Begin to relate addition to combining two groups of objects and subtraction to 'taking away' (Level 1)	Use the vocabulary related to addition and subtraction and symbols to describe and record addition and subtraction number sentences (Level 2)	Add or subtract mentally a one-digit number or a multiple of 10 to or from any two-digit number; use practical and informal written methods to add and subtract two-digit numbers (Level 2)	Add or subtract mentally combinations of one- and two-digit numbers (Level 3)
In practical activities and discussion begin to use the vocabulary involved in adding and subtracting (Level 1)		Use the symbols +, −, ×, ÷ and = to record and interpret number sentences involving all four operations; calculate the value of an unknown in a number sentence (Level 2)	
Understanding shape (AT3)			
Use language such as 'circle' or 'bigger' to describe the shape and size of solids and flat shapes (Level 1)	Visualise and name common 2-D shapes and 3-D solids and describe their features; use them to make patterns, pictures and models (Level 1)	Visualise common 2-D shapes and 3-D solids; identify shapes from pictures of them in different positions and orientations; sort, make and describe shapes, referring to their properties (Level 2)	Draw and complete shapes with reflective symmetry and draw the reflection of a shape in a mirror line along one side (Level 3)
Use everyday words to describe position (Level 1)			
Measuring (AT3)			
Use language such as 'greater', 'smaller', 'heavier' or 'lighter' to compare quantities (Level 1)	Estimate, measure, weigh and compare objects, choosing and using suitable uniform non-standard or standard units and measuring instruments, e.g. a lever balance, metre stick or measuring jug (Level 2)	Use units of time (seconds, minutes, hours, days) and know the relationships between them; read the time to the quarter hour; identify time intervals, including those that cross the hour (Level 2)	Read, to the nearest division and half-division, scales that are numbered or partially numbered; use the information to measure and draw to a suitable degree of accuracy (Level 3)
Handling data (AT4)			
	Answer a question by recording information in lists and tables; present outcomes using practical resources, pictures, block graphs or pictograms (Level 2)	Use lists, tables and diagrams to sort objects; explain choices using appropriate language, including not (Level 2)	Use Venn diagrams or Carroll diagrams to sort data and objects using more than one criterion (Level 3)

NOTES: **Using and applying mathematics (AT1)** is incorporated throughout

Record-keeping format 10 Individual child's record of the end-of-year expectations   Name: _____

Year 4	Year 5	Year 6	Year 6 progression to Year 7
Counting and understanding number (AT2)			
Use diagrams to identify equivalent fractions; interpret mixed numbers and position them on a number line (Level 3)	Explain what each digit represents in whole numbers and decimals with up to two places, and partition, round and order these numbers (Level 3)	Express one quantity as a percentage of another; find equivalent percentages, decimals and fractions (Level 4)	Use ratio notation, reduce a ratio to its simplest form and divide a quantity into two parts in a given ratio; solve simple problems involving ratio and direct proportion (Level 5)
Knowing and using number facts (AT2)			
Derive and recall multiplication facts up to 10 × 10, the corresponding division facts and multiples of numbers to 10 up to the tenth multiple (Level 4)	Use knowledge of place value and addition and subtraction of two-digit numbers to derive sums and differences, doubles and halves of decimals (Level 4)	Use knowledge of place value and multiplication facts to 10 × 10 to derive related multiplication and division facts involving decimals (Level 4)	Make and justify estimates and approximations to calculations (Level 5)
Calculating (AT2)			
Add or subtract mentally pairs of two-digit whole numbers (Level 3)	Use efficient written methods to add and subtract whole numbers and decimals with up to two places (Level 4)	Use efficient written methods to add and subtract integers and decimals, to multiply and divide integers and decimals by a one-digit integer, and to multiply two- and three-digit integers by a two-digit integer (Level 4)	Use bracket keys and the memory of a calculator to carry out calculations with more than one step; use the square-root key (Level 5)
Develop and use written methods to record, support and explain multiplication and division of two-digit numbers by a one-digit number, including division with remainders (Level 3)			
Understanding shape (AT3)			
Know that angles are measured in degrees and that one whole turn is 360°; compare and order angles less than 180° (Level 3)	Read and plot co-ordinates in the first quadrant; recognise parallel and perpendicular lines in grids and shapes; use a set-square and ruler to draw shapes with perpendicular or parallel sides (Level 4)	Visualise and draw on grids of different types where a shape will be after reflection, after translations, or after rotation through 90° or 180° about its centre or one of its vertices (Level 4)	Know the sum of angles on a straight line, in a triangle and at a point, and recognise vertically opposite angles (Level 5)
Measuring (AT3)			
Choose and use standard metric units and their abbreviations when estimating, measuring and recording length, weight and capacity; know the meaning of kilo, centi and milli and, where appropriate, use decimal notation to record measurements (Level 3)	Draw and measure lines to the nearest millimetre; measure and calculate the perimeter of regular and irregular polygons; use the formula for the area of a rectangle to calculate its area (Level 4)	Select and use standard metric units of measure and convert between units using decimals to two places (Level 4)	Solve problems by measuring, estimating and calculating; measure and calculate using imperial units still in everyday use; know their approximate metric values (Level 5)
Handling data (AT4)			
Answer a question by identifying what data to collect; organise, present, analyse and interpret the data in tables, diagrams, tally charts, pictograms and bar charts, using ICT where appropriate (Level 3)	Construct frequency tables, pictograms and bar and line graphs to represent the frequencies of events and changes over time (Level 4)	Solve problems by collecting, selecting, processing, presenting and interpreting data, using ICT where appropriate; draw conclusions and identify further questions to ask (Level 4)	Understand and use the probability scale from 0 to 1; find and justify probabilities based on equally likely outcomes in simple contexts (Level 5)

NOTES: **Using and applying mathematics (AT1)** is incorporated throughout

	Foundation Stage	Year 1	Year 2	Year 3	Year 4	Year 5	Year 6
End-of-year National Expectations	1b	1a (2c)	2b	2a (3c)	3b	3a (4c)	4b

Word problem cards

1. A water company has 10 tanks each containing 2·5 million litres of water. What is the total capacity of the 10 tanks?

2. A tin holds 2 litres of paint. One tin contains enough paint to cover an area of 25 m². If the total surface area of the classroom is 120 m², how many tins are needed to paint the classroom?

3. It takes Justin 395 seconds to run around the school playground. It takes Leroy 325 seconds to run the same distance. How much faster is Leroy in minutes and seconds?

4. Anita is making new curtains. She buys a length of material 3·76 m long. She cuts it into two equal lengths. What is the length of each piece?

5. A one litre bottle of orange juice contains 30% pure juice. How many millilitres of pure juice are in one bottle?

6. In the triple jump, Henry hopped 2·9 metres, stepped 2·1 metres and jumped 3·3 metres. How long was his triple jump?

7. It costs 60p to park for 20 minutes in the High Street. Teresa parks for 1 hour and 40 minutes. How much must she pay?

8. The Ocean Princess holds 2600 passengers. There are 2 restaurants, each with 2 sittings for each meal. If the passengers are divided equally between the 2 restaurants and the 2 sittings, how many passengers are in each sitting?

9. In the children's section of a local library there are 8294 fiction books and 6468 non-fiction books. On the 29th May, 1064 children's books were out on loan. How many children's books were still in the library?

10. Garden chairs cost £12 each. However, if you buy more than 6 there is a 20% reduction in the total price. If David buys 8 chairs how much will he have to pay?

11. Freda is buying a new carpet for her living room. The carpet costs £8 per square metre. If the room is 6·5 m wide by 8 m long, how much does the carpet cost?

12. A box contains 8 jars of apricot jam. Each jar weighs 370 g. What is the weight of the 8 jars?

13. When Peter is waking up in London at 07:30 on Monday morning, his grandmother in Australia is going to bed at 18:30 on Monday evening. What is the time difference between London and Australia?

14. Mr. and Mrs. Sims went on a holiday to Australia. They spent £2684 on flights, £1587 on accommodation and £1652 on other expenses. How much did their holiday cost them altogether?

15. The normal price of a car is £9200. In a sale Mrs. Freeman saves £1840. What is the percentage saving she made?

16. In 2000 Kevin bought a house for £58 000. By 2007 its value had increased by 32%. What was the new value of the house?

17. Cheese is delivered to the supermarket in 2·5 kg boxes. A box contains 20 trays of cheese each tray weighing the same amount. If 8 trays of cheese are sold, what is the weight of cheese left in the box?

18. Last year the number of spectators at a car rally was 12 600. This year the number of spectators rose by 5%. How many spectators were at the car rally this year?

19. The Red Rose Café makes 8 fruit pies. Each pie is cut into eighths to sell as slices. Altogether the café sells $5\frac{3}{4}$ pies. How many slices of the 8 pies are left?

20. I am thinking of a number. If you subtract me from 10 and add 0·8, you get 6. What am I?

Puzzles 1

$$20 = 4^2 + 2^2$$

20 can be calculated by finding the sum of two square numbers.

Investigate which numbers from 1 to 50 can be made by calculating the sum or difference between two square numbers.

Write about what you discover in words and symbols.

● Each missing number in the wall is made by finding the sum of the two numbers below it.

● Look at the walls below. The letters v, w, x, y and z stand for numbers.

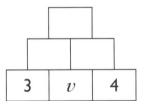
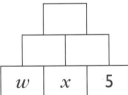

● Complete each wall.

● Now investigate the relationship between the top number and the three numbers in the bottom row.

● Write the relationship in words and symbols.

Collins
New
Primary
Maths

Puzzles 2

The sum of the interior angles of a triangle total 180°.

The sum of the interior angles of a quadrilateral total 360°.

The sum of the interior angles of a pentagon total 540°.

What is the sum of the interior angles of each of the following shapes?
- hexagon
- heptagon
- octagon

Write about what you notice.

Write a formula for the sum of angles in regular polygons.

- Look at this sequence of triangles.

- Continue the sequence for 4 more triangles.

- Investigate the relationship between the total number of dots in each new triangle.

- Write the relationship in words and symbols.

- Now use your formula to write the sequence of triangular numbers until you pass 100.

- What do you notice about the sum of two consecutive triangular numbers?

- What do you notice about the difference between two consecutive triangular numbers?

© Collins New Primary Maths

Temperatures

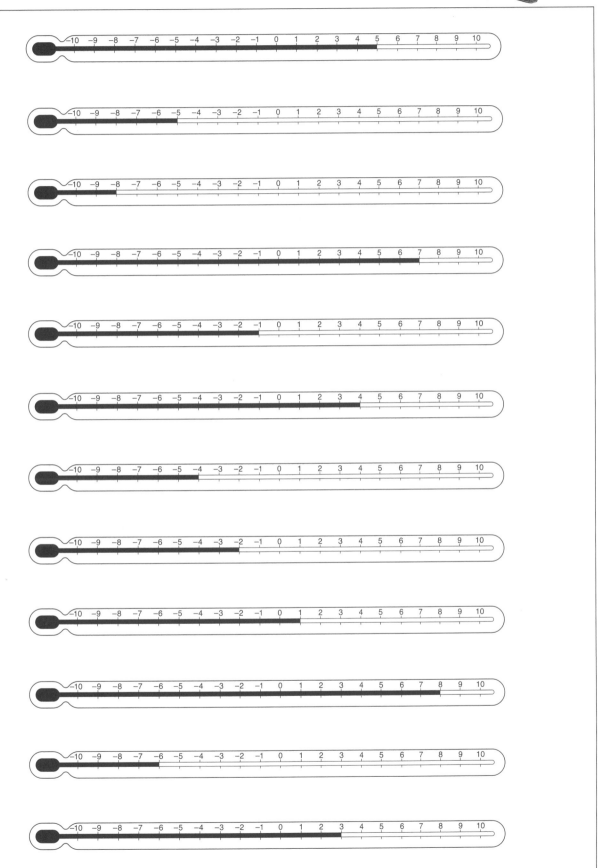

Collins New Primary Maths

Decimal numbers

1·24	5·702	9·3	3·62	0·08	0·11
6·371	4·25	0·98	7·61	8·7	3·348
2·35	1·34	7·38	5·59	2·03	3·78
8·69	5·06	1·142	3·912	7·146	5·46
3·42	9·81	8·701	0·1	4·45	1·01
1·42	4·52	7·83	1·43	2·38	5·101
8·73	6·317	0·8	5·802	1·239	7·29
0·89	4·6	8·710	2·83	7·461	3·78
5·6	1·1	9·072	5·64	7·39	9·078
6·731	8·37	6·137	7·93	7·416	7·92

Reducing fractions

9	$\frac{120}{360}$	$\frac{30}{54}$	$\frac{15}{20}$	$\frac{18}{27}$	$\frac{8}{16}$	$\frac{12}{44}$	$\frac{2}{4}$	$\frac{16}{36}$	$\frac{8}{12}$	$\frac{5}{55}$
8	$\frac{4}{18}$	$\frac{60}{100}$	$\frac{6}{42}$	$\frac{3}{21}$	$\frac{16}{64}$	$\frac{3}{6}$	$\frac{6}{9}$	$\frac{15}{35}$	$\frac{40}{60}$	$\frac{3}{9}$
7	$\frac{2}{10}$	$\frac{20}{32}$	$\frac{8}{20}$	$\frac{6}{16}$	$\frac{2}{6}$	$\frac{12}{14}$	$\frac{6}{15}$	$\frac{6}{8}$	$\frac{4}{12}$	$\frac{2}{8}$
6	$\frac{14}{16}$	$\frac{8}{14}$	$\frac{7}{14}$	$\frac{3}{18}$	$\frac{6}{10}$	$\frac{2}{14}$	$\frac{4}{6}$	$\frac{9}{21}$	$\frac{5}{10}$	$\frac{25}{30}$
5	$\frac{3}{33}$	$\frac{14}{49}$	$\frac{15}{18}$	$\frac{18}{54}$	$\frac{2}{18}$	$\frac{5}{20}$	$\frac{4}{100}$	$\frac{2}{20}$	$\frac{10}{15}$	$\frac{50}{100}$
4	$\frac{10}{18}$	$\frac{6}{20}$	$\frac{4}{24}$	$\frac{6}{36}$	$\frac{8}{18}$	$\frac{10}{12}$	$\frac{6}{18}$	$\frac{20}{30}$	$\frac{12}{27}$	$\frac{5}{30}$
3	$\frac{16}{18}$	$\frac{4}{16}$	$\frac{21}{30}$	$\frac{40}{800}$	$\frac{6}{27}$	$\frac{4}{8}$	$\frac{6}{12}$	$\frac{20}{24}$	$\frac{9}{12}$	$\frac{10}{14}$
2	$\frac{14}{20}$	$\frac{72}{720}$	$\frac{12}{20}$	$\frac{2}{16}$	$\frac{12}{16}$	$\frac{8}{24}$	$\frac{9}{180}$	$\frac{3}{24}$	$\frac{3}{15}$	$\frac{9}{15}$
1	$\frac{21}{24}$	$\frac{4}{14}$	$\frac{9}{30}$	$\frac{27}{99}$	$\frac{10}{16}$	$\frac{18}{20}$	$\frac{6}{24}$	$\frac{21}{27}$	$\frac{27}{30}$	$\frac{5}{15}$
0	$\frac{3}{12}$	$\frac{12}{40}$	$\frac{12}{15}$	$\frac{9}{24}$	$\frac{15}{24}$	$\frac{6}{21}$	$\frac{3}{30}$	$\frac{6}{14}$	$\frac{14}{18}$	$\frac{10}{20}$
	0	1	2	3	4	5	6	7	8	9

Collins
New
Primary
Maths

Fractions, decimals and percentages cards

$\dfrac{1}{2}$	$\dfrac{1}{10}$	$\dfrac{3}{5}$	$\dfrac{9}{10}$
$\dfrac{1}{100}$	$\dfrac{1}{25}$	0·25	0·4
0·3	0·8	0·02	0·15
0·62	8%	20%	70%
75%	5%	12%	24%

Collins New Primary Maths

Ratio and proportion word problem cards

1. A leg of lamb should be cooked for 40 minutes for every kilogram. How long does it take to cook a 3 kilogram leg of lamb?

2. Four chocolates cost 20p pence altogether. How much do twelve chocolates cost?

3. Larry is making squash. The finished drink should be $\frac{1}{3}$ squash and $\frac{2}{3}$ water. Larry puts 100 ml of squash in a glass. How much water should he put with it?

4. In a box containing 20 Christmas decorations, one in every five decorations is broken. How many decorations are broken?

5. Grapes cost £1.20 for 100 grams. What is the cost of 350 grams of grapes?

6. Two books cost eighty pence. How much do three books cost?

7. A local supermarket is selling 12 cans of drink for £3.60. What is the cost of 8 cans of drink?

8. 500 g of olives are divided so that Jamie gets three times as much as Gordon. How much does Gordon get?

9. Two letters have a total weight of 180 grams. One letter weighs twice as much as the other. Write the weight of the heavier letter?

10. Pumpkin seeds cost 80p for 100 grams. Leroy pays £2 for a bag of pumpkin seeds. How many grams of pumpkin seeds does he get?

11. In the Northview Football Club, 3 in every 5 members are boys. What percentage of the members are boys?

12. Six cakes cost £2.40. How much do ten cakes cost?

13. There are 15 children in a beginners tennis group. For every 3 girls there are 2 boys. How many are girls?

14. In an after school chess club there are 3 boys for every 2 girls. There are 40 children altogether in the club. How many girls are there?

15. A recipe for 3 people needs 75 g of flour.
a) How much flour do you need for 2 people?
b) How much flour do you need for 7 people?

16. In a dance there are 3 boys and 2 girls in every line. 48 boys take part in the dance.
a) How many girls take part?
b) For a different dance there are 50 children. How many boys are there?

17.

Winston School's donations to Winston Hospital			
Rec: £60	Year 1: £80	Year 2: £120	Year 3: £100
Year 4: £140		Year 5: £200	Year 6: £100

a) What proportion of the total money was donated by Year 5?
b) What was the ratio of money donated by Year 6 compared to Year 4?

18.

Fruit cocktail	
1 part lime juice	5 parts fruit purée
2 parts sugar syrup	4 parts soda water

Lee makes the fruit cocktail using 200 ml of sugar syrup.
a) How much of each of the other ingredients does he use?
b) How much drink does he make altogether?

19. A 4·2 kg cake consists of 3 parts in 7 of flour and 1 part in 3 of sugar. How much do the other ingredients in the cake weigh?

20. It costs £7.20 to wash and dry 5 kg of laundry at a launderette. The ratio of the cost of washing to the cost of drying is 5 : 3.
a) How much does the washing cost?
b) How much would it cost to dry 7 kg of laundry?

Collins New Primary Maths

Calculating with decimals

2.9   5.8   7.7   6.4   4.1   3.7   8.5   7.2   9.3   1.6

 0.1

 0.2

0.5

 0.8

0.7

 0.3

 0.4

 0.9

0.6

 4.2

2.1

3.2

4.5

2.4

 1.8

 3.5

 1.2

 6.3

 4.8

Collins New Primary Maths

Estimate, calculate and check

Estimation	Calculation	Check

Collins New Primary Maths

Two-digit number cards

14	16	23	25
33	39	40	47
51	57	62	64
66	71	74	78
82	87	95	98

© Collins New Primary Maths

Three-digit number cards

164	186	224	293
305	332	457	479
515	548	613	662
681	739	756	795
846	867	908	974

Collins New Primary Maths

Four-digit number cards

1292	1615	2302	2791
3071	3862	4186	4976
5560	5838	6057	6257
7113	7390	7605	8284
8429	8744	9443	9538

Collins
New
Primary
Maths

Decimal cards – tenths

0·7	1·5	3·6	4·3
7·2	9·4	25·1	32·7
37·9	52·3	68·4	83·8
117·2	304·4	457·6	645·1
648·9	703·5	891·7	955·8

Collins
New
Primary
Maths

Decimal cards – hundredths

0·09	1·43	2·31	3·51
4·35	5·82	6·82	7·97
8·22	9·68	14·68	26·74
34·16	48·73	52·46	60·05
65·53	71·97	88·19	93·24

Relating fractions to multiplication and division

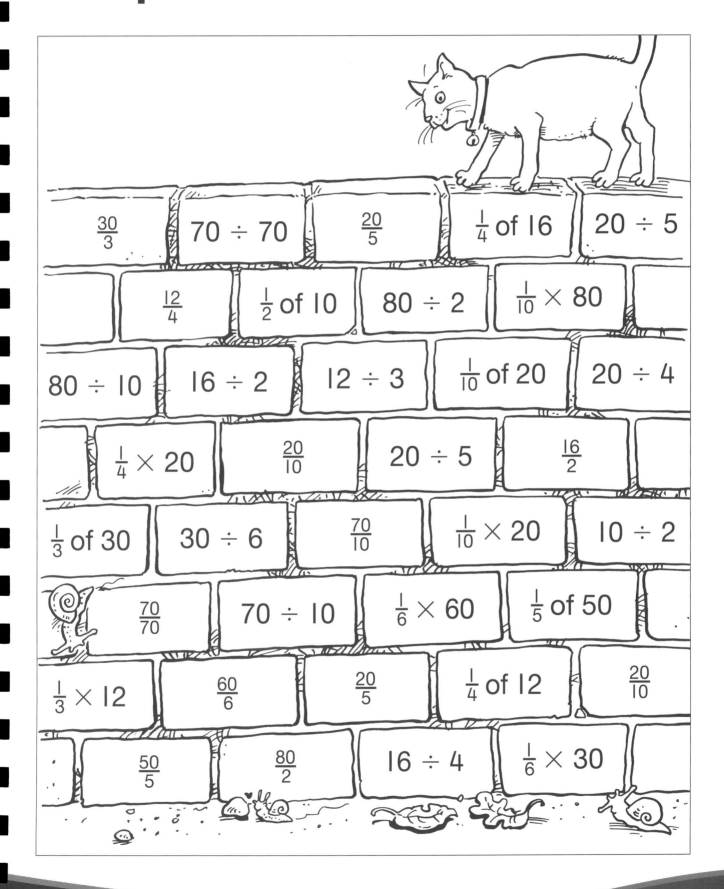

$\frac{30}{3}$

$70 \div 70$

$\frac{20}{5}$

$\frac{1}{4}$ of 16

$20 \div 5$

$\frac{12}{4}$

$\frac{1}{2}$ of 10

$80 \div 2$

$\frac{1}{10} \times 80$

$80 \div 10$

$16 \div 2$

$12 \div 3$

$\frac{1}{10}$ of 20

$20 \div 4$

$\frac{1}{4} \times 20$

$\frac{20}{10}$

$20 \div 5$

$\frac{16}{2}$

$\frac{1}{3}$ of 30

$30 \div 6$

$\frac{70}{10}$

$\frac{1}{10} \times 20$

$10 \div 2$

$\frac{70}{70}$

$70 \div 10$

$\frac{1}{6} \times 60$

$\frac{1}{5}$ of 50

$\frac{1}{3} \times 12$

$\frac{60}{6}$

$\frac{20}{5}$

$\frac{1}{4}$ of 12

$\frac{20}{10}$

$\frac{50}{5}$

$\frac{80}{2}$

$16 \div 4$

$\frac{1}{6} \times 30$

Collins New Primary Maths

Finding fractions of numbers

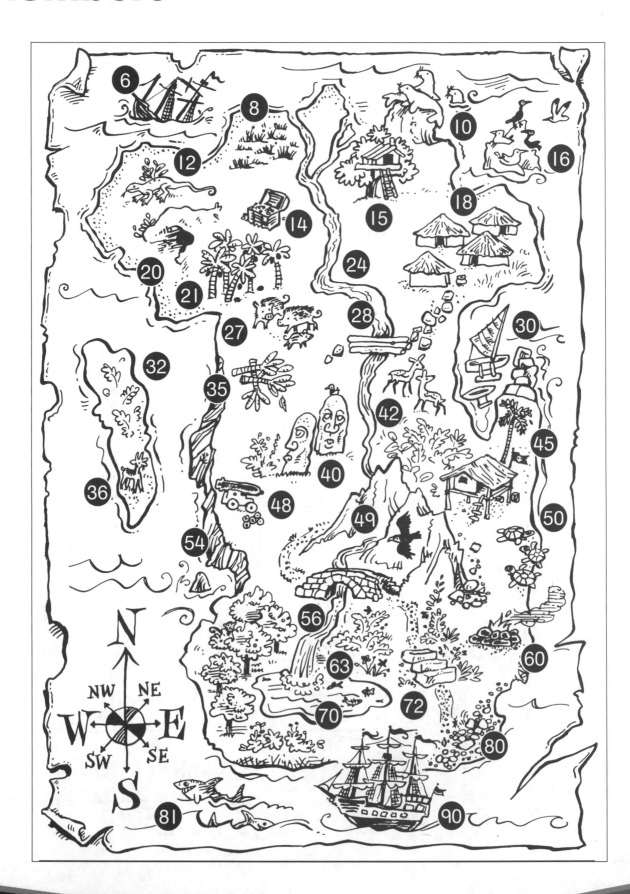

Percentages

1% 10% 25% 50% 75%

2% 5% 20% 30% 40%

45% 60% 70% 80% 90%

12% 18% 46% 54% 62%

10 20 30 40

50 60 70 80

90 100 110 120

130 140 150 160

170 180 190 200

300 400 500 600

Collins New Primary Maths

2-D shapes

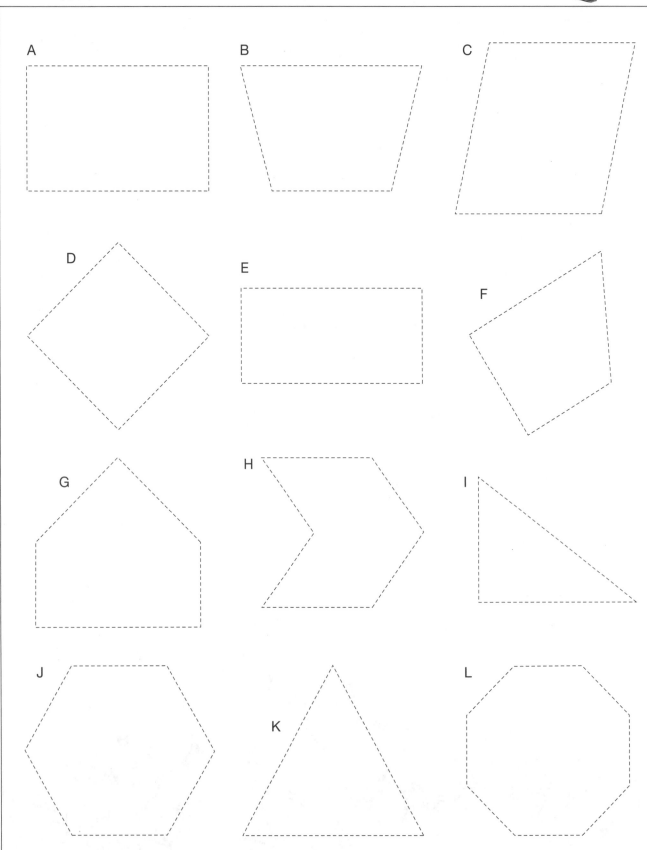

A

B

C

D

E

F

G

H

I

J

K

L

Collins New Primary Maths

Reflect, translate and rotate

Grid 1

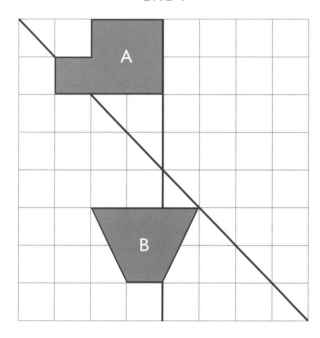

Grid 2

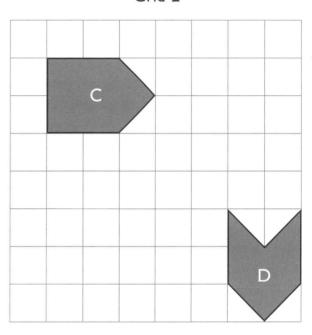

Grid 3

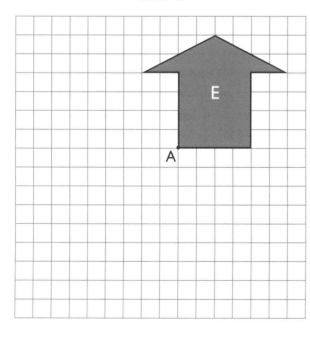

Grid 4

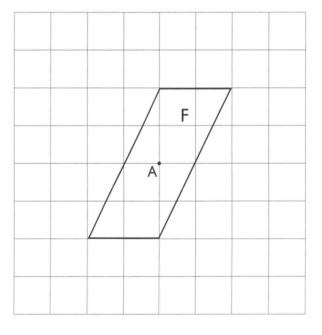

First quadrant co-ordinates

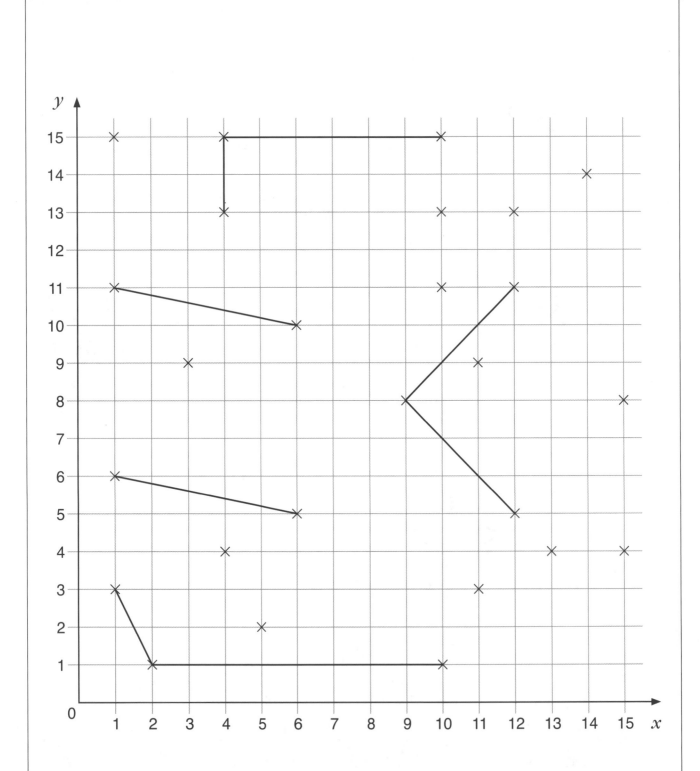

Measuring angle cards

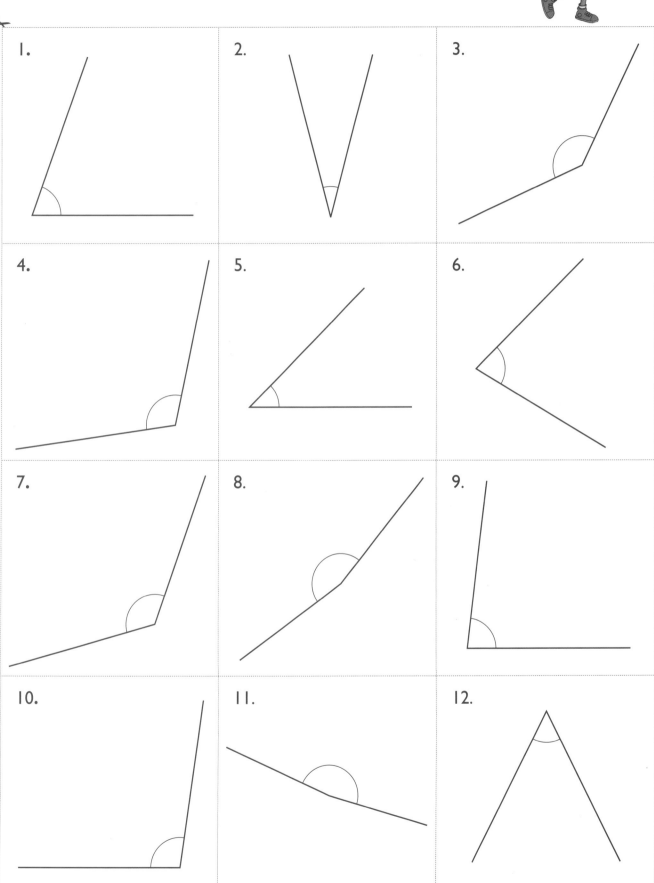

1.

2.

3.

4.

5.

6.

7.

8.

9.

10.

11.

12.

Collins
New
Primary
Maths

Angles in a triangle and around a point cards

1.
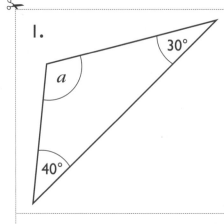
30°
a
40°

2.
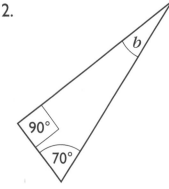
b
90°
70°

3.
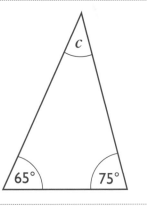
c
65°   75°

4.
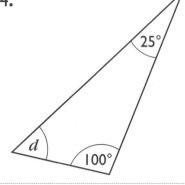
25°
d
100°

5.
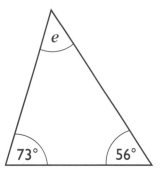
e
73°   56°

6.

24°
f   92°

7.
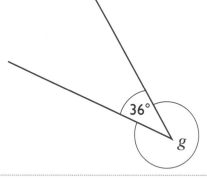
36°
g

8.
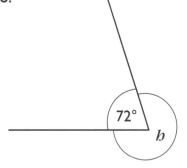
72°
h

9.
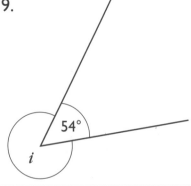
54°
i

10.
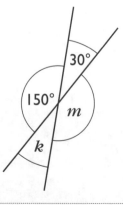
30°
150°
m
k

11.
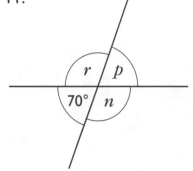
r *p*
70° *n*

12.
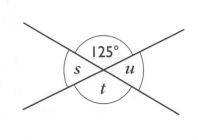
125°
s   *u*
t

Collins
New
Primary
Maths

Reading and interpreting scales – Mass

Reading and interpreting scales – Capacity

1.

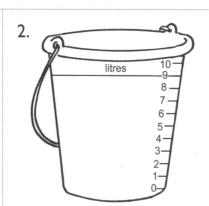

2.

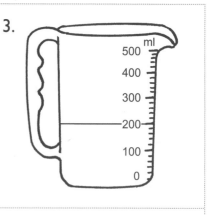

3.

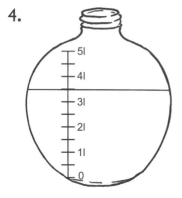

4.

5.

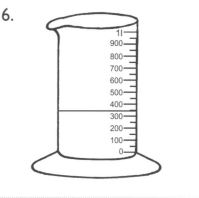

6.

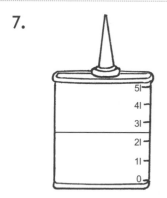

7.

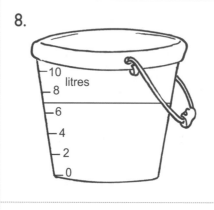

8.

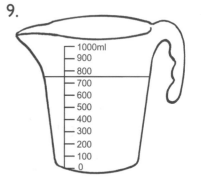

9.

10.

11.

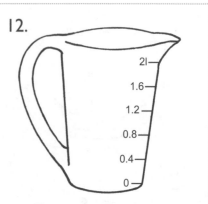

12.

Collins New Primary Maths

Perimeter and area cards 1

1.

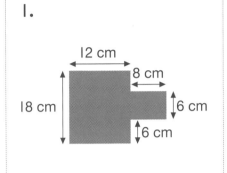

12 cm
8 cm
18 cm
6 cm
6 cm

2.

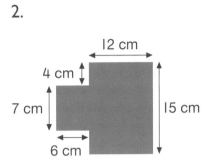

12 cm
4 cm
7 cm
15 cm
6 cm

3.

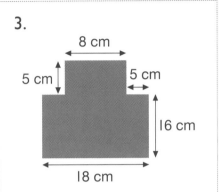

8 cm
5 cm
5 cm
16 cm
18 cm

4.

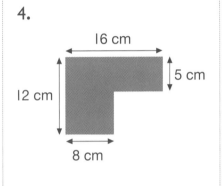

16 cm
12 cm
5 cm
8 cm

5.

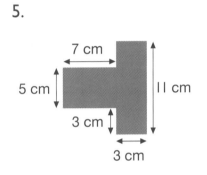

7 cm
5 cm
11 cm
3 cm
3 cm

6.

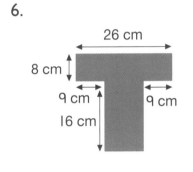

26 cm
8 cm
9 cm
9 cm
16 cm

7.

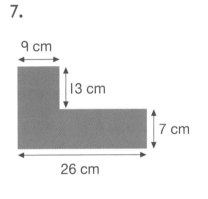

9 cm
13 cm
7 cm
26 cm

8.

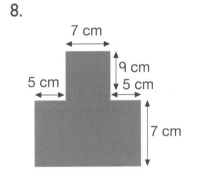

7 cm
5 cm
9 cm
5 cm
7 cm

9.

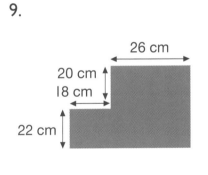

26 cm
20 cm
18 cm
22 cm

10.

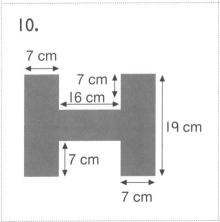

7 cm
7 cm
16 cm
19 cm
7 cm
7 cm

11.

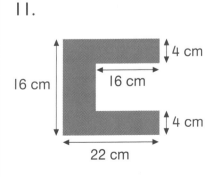

4 cm
16 cm
16 cm
4 cm
22 cm

12.

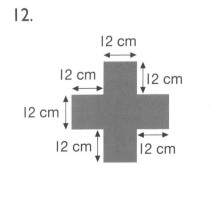

12 cm
12 cm
12 cm
12 cm
12 cm
12 cm

Collins
New
Primary
Maths

Perimeter and area cards 2

1.

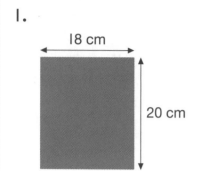

18 cm

20 cm

2.

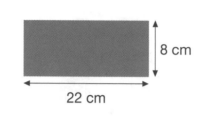

8 cm

22 cm

3.

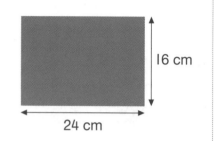

16 cm

24 cm

4.

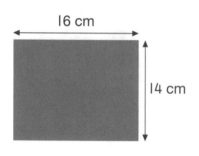

16 cm

14 cm

5.

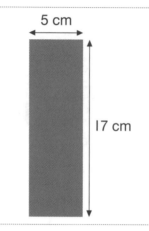

5 cm

17 cm

6.

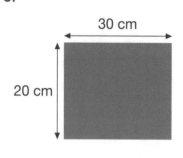

30 cm

20 cm

7.

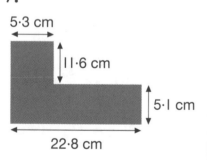

5·3 cm

11·6 cm

5·1 cm

22·8 cm

8.

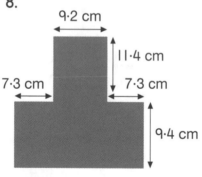

9·2 cm

11·4 cm

7·3 cm   7·3 cm

9·4 cm

9.

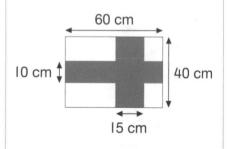

60 cm

10 cm

40 cm

15 cm

10.

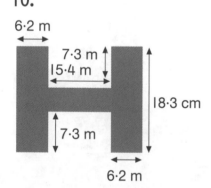

6·2 m

7·3 m

15·4 m

18·3 cm

7·3 m

6·2 m

11.

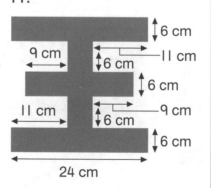

6 cm

9 cm

11 cm

6 cm

6 cm

11 cm

9 cm

6 cm

24 cm

6 cm

12.

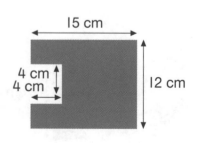

15 cm

4 cm

4 cm

12 cm

Collins
New
Primary
Maths

Probability

good chance

poor chance

certain

even chance

impossible

Collecting, selecting and organising data

Answer the following question by collecting, selecting and organising relevant data.

Things to think about

● How are you going to collect your data?
Perhaps using a tally chart or frequency table.

● How are you going to present your data?
Perhaps in a table or graph.

My initial ideas

How I am going to collect the data

How I am going to organise the data

Now do it!

What I found out

What else I could find out about this topic

Collins
New
Primary
Maths

Handling data 1

Paul the Baker – Amount of money taken on 16th July

Amount of money taken (£) vs Time of day

Davis Dry Cleaners – Items brought in for cleaning on 16th July

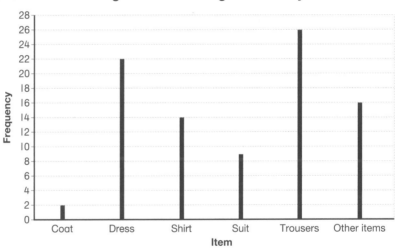

Frequency vs Item

Monthly average temperature in London

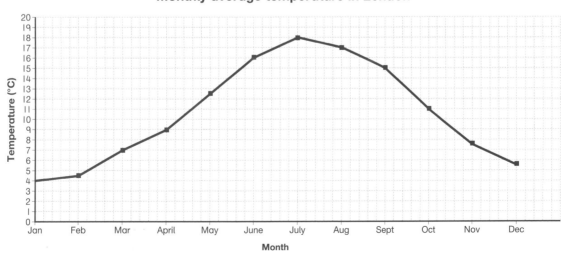

Temperature (°C) vs Month

Collins New Primary Maths

Handling data 2

Pounds to euros conversion graph

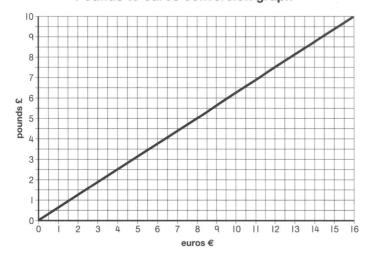

pounds £ (y-axis)
euros € (x-axis)

Travel Bug Customers Favourite Holiday Destinations

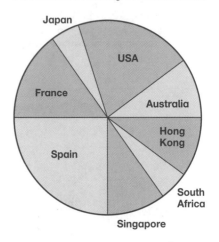

Japan, USA, France, Australia, Hong Kong, Spain, South Africa, Singapore

Road distances in Italy

km	Bologna	Florence	Milan	Naples	Pisa	Rome	Siena	Venice	Verona
Bologna		105	210	590	180	380	170	155	140
Florence	105		300	490	95	280	70	255	230
Milan	210	300		785	275	575	365	270	160
Naples	590	490	785		585	220	440	740	715
Pisa	180	95	275	585		375	110	330	280
Rome	380	280	575	220	375		230	530	505
Siena	170	70	365	440	110	230		320	295
Venice	155	255	270	740	330	530	320		115
Verona	140	230	160	715	280	505	295	115	

Collins New Primary Maths